Undergraduate Lecture Notes in Physics

D1262865

Undergraduate Lecture Notes in Physics (ULNP) publishes authoritative texts covering topics throughout pure and applied physics. Each title in the series is suitable as a basis for undergraduate instruction, typically containing practice problems, worked examples, chapter summaries, and suggestions for further reading.

ULNP titles must provide at least one of the following:

- An exceptionally clear and concise treatment of a standard undergraduate subject.
- A solid undergraduate-level introduction to a graduate, advanced, or non-standard subject.
- A novel perspective or an unusual approach to teaching a subject.

ULNP especially encourages new, original, and idiosyncratic approaches to physics teaching at the undergraduate level.

The purpose of ULNP is to provide intriguing, absorbing books that will continue to be the reader's preferred reference throughout their academic career.

Series editors

Neil Ashby
Professor Emeritus, University of Colorado, Boulder, CO, USA

William Brantley
Professor, Furman University, Greenville, SC, USA

Michael Fowler
Professor, University of Virginia, Charlottesville, VA, USA

Morten Hjorth-Jensen
Oslo, Norway

Michael Inglis
Professor, SUNY Suffolk County Community College, Selden, NY, USA

Heinz Klose
Professor Emeritus, Humboldt University Berlin, Germany

Helmy Sherif
Professor, University of Alberta, Edmonton, AB, Canada

More information about this series at http://www.springer.com/series/8917

Tilman Butz

Fourier Transformation
for Pedestrians

Second Edition

 Springer

Tilman Butz
Faculty of Physics and Earth Science
University of Leipzig
Leipzig
Germany

ISSN 2192-4791   ISSN 2192-4805   (electronic)
Undergraduate Lecture Notes in Physics
ISBN 978-3-319-16984-2   ISBN 978-3-319-16985-9   (eBook)
DOI 10.1007/978-3-319-16985-9

Library of Congress Control Number: 2015935732

Springer Cham Heidelberg New York Dordrecht London

© Springer International Publishing Switzerland 2006, 2015
This work is subject to copyright. All rights are reserved by the Publisher, whether the whole or part of the material is concerned, specifically the rights of translation, reprinting, reuse of illustrations, recitation, broadcasting, reproduction on microfilms or in any other physical way, and transmission or information storage and retrieval, electronic adaptation, computer software, or by similar or dissimilar methodology now known or hereafter developed.
The use of general descriptive names, registered names, trademarks, service marks, etc. in this publication does not imply, even in the absence of a specific statement, that such names are exempt from the relevant protective laws and regulations and therefore free for general use.
The publisher, the authors and the editors are safe to assume that the advice and information in this book are believed to be true and accurate at the date of publication. Neither the publisher nor the authors or the editors give a warranty, express or implied, with respect to the material contained herein or for any errors or omissions that may have been made.

Springer International Publishing AG Switzerland is part of Springer Science+Business Media
(www.springer.com)

To Renate, Raphaela, and Florentin

Preface

Fourier[1] Transformation for Pedestrians. For *pedestrians*? Harry J. Lipkin's famous "Beta-decay for Pedestrians" [1], was an inspiration to me, so that's why. Harry's book explains physical problems as complicated as helicity and parity violation to "pedestrians" in an easy to understand way. Discrete Fourier transformation, by contrast, only requires elementary algebra, something any student should be familiar with. As the algorithm[2] is a linear one, this should present no pitfalls and should be as "easy as pie". In spite of that, stubborn prejudices prevail, as far as Fourier transformations are concerned, viz., that information could get lost or that you could end up trusting a hoax; anyway, who would trust something that is all done with "smoke and mirrors". The above prejudices are often caused by negative experiences, gained through improper use of ready-made Fourier transformation programs or hardware.

This book is for all who, being laypersons—or pedestrians—are looking for a gentle and also humorous introduction to the application of Fourier transformation, without hitting too much theory, proofs of existence, and similar things. It is appropriate for science students at technical colleges and universities, but also for "mere" computer-freaks. It is also quite adequate for students of engineering and all practical people working with Fourier transformations. Basic knowledge of integration, however, is recommended.

If this book can help to avoid prejudices or even do away with them, writing it has been well worthwhile. Here we show how things "work". Generally we discuss the Fourier transformation in one dimension only. Chapter 1 introduces Fourier series and, as part and parcel, important statements and theorems that will guide us through the whole book. As is appropriate for pedestrians, we also cover all the "pits and pitfalls" on the way. Chapter 2 covers continuous Fourier transformations in great detail. Window functions are dealt with in Chap. 3 in more detail, as

---

[1]Jean Baptiste Joseph Fourier (1768–1830), French mathematician and physicist.

[2]Integration and differentiation are linear operators. This is quite obvious in the discrete version (Chap. 4) and is, of course, also valid when passing on to the continuous form.

understanding them is essential to avoid the disappointment caused by false expectations. Chapter 4 is about discrete Fourier transformations, with special regard to the Cooley–Tukey algorithm (Fast Fourier Transform, FFT). Finally, Chap. 5 will introduce some useful examples for the filtering effects of simple algorithms. From the host of available material we only pick items that are relevant to the recording and preprocessing of data, items that are often used without even thinking about them. This book started as a manuscript for lectures at the Technical University of Munich and at the University of Leipzig. That is why it is very much a textbook and contains many worked examples—to be redone "manually"—as well as plenty of illustrations. To show that a textbook (originally) written in German can also be amusing and humorous, was my genuine concern, because dedication and assiduity of their own are quite inclined to stifle creativity and imagination. It should also be fun and boost our innate urge to play. The two books "Applications of Discrete and Continuous Fourier Analysis" [2] and "Theory of Discrete and Continuous Fourier Analysis" [3] had considerable influence on the makeup and content of this book, and are to be recommended as additional reading for those "keen on theory".

This English edition is based on the third, enlarged edition in German [4]. In contrast to this German edition, there are now problems at the end of each chapter. They should be worked out before going to the next chapter. However, I prefer the word "playground" because you are allowed to go straight to the solutions, compiled in the Appendix, should your impatience get the better of you. In case you've read the German original, there I apologized for using many new-German words, such as "sampeln" or "wrappen"; I won't do that here, to the contrary, they come in very handy and make the translator's job (even) easier. Many thanks to Mrs. U. Seibt and Mrs. K. Schandert, as well as to Dr. T. Reinert, Dr. T. Soldner and especially to Mr. H. Gödel (Dipl.-Phys.) for the hard work involved in turning a manuscript into a book. Mr. St. Jankuhn (Dipl.-Phys.) did an excellent job in proofreading and computer acrobatics.

Last but not least, special thanks go to the translator who managed to convert the informal German style into an informal ("downunder") English style.

Recommendations, queries, and proposals for change are welcome. Have fun while reading, playing, and learning.

Leipzig   Tilman Butz
April 2005

Preface of the Translator

More than a few moons ago I read two books about Richard Feynman's life, and that has made a lasting impression. When Tilman Butz asked me if I could translate his "Fourier Transformation for Pedestrians", I leapt at the chance—my way of getting a bit more into science. During the rather mechanical process of translating the German original, within its TEX-framework, I made sure I enjoyed the bits for the pedestrians, mere mortals like myself. Of course I am biased, I have known the author for many years—after all he is my brother.

Hamilton, New Zealand
2004

Thomas-Severin Butz

Preface for the Second Enlarged Edition

The second enlarged edition is based on the first edition with a focus on applications in signal analysis and processing. In a digital world, the discrete Fourier transformation plays a very important role. However, in order to avoid pitfalls, it is strongly recommended to learn about Fourier series and continuous Fourier transformation first. Two new chapters were added to the first edition for pedestrians that like to go a little further: the first deals with data streams and fractional delays, a topic which is important in a variety of fields ever since the development of fast digitizers; the second gives an introduction to tomography with focus on a common image reconstruction algorithm, the back-projection of filtered projections. Here, we shall use the spatial coordinate x and the "angular wave number" k instead of time t and angular frequency ω. Both topics are intimately related to Fourier transformation and deal with modern applications. Occasionally, series and integrals are needed which are beyond elementary calculus. For those who have access to a good library, references are given to verify the results. Those who have access to Mathematica [5] will prefer to use this tool instead. They will miss the admiration of human ingenuity of the pre-computer era but hopefully will admire Mathematica. Since the focus of this book is on signal processing, important issues like proofs of existence, convergence of infinite series, permutability of integration and differentiation, and integrability of functions are not addressed. Those pedestrians who want to step a little deeper into mathematics are encouraged to do so. A large number of typos and errors have been corrected (unfortunately you never get hold of all) and critical comments have been taken care of. It is a pleasure to thank Mr. St. Jankuhn (Dipl.-Phys.) for his excellent work to complete the second edition and Mr. M. Jäger (Dipl.-Inf.) for fruitful discussions on fractional delays. Recommendations, queries and proposals for change are always welcome. Have fun while reading, playing and learning.

Leipzig   Tilman Butz

Contents

Introduction

One of the general tasks in science and engineering is to record measured signals and get them to tell us their "secrets" (information). Here we are mainly interested in signals varying over time. They may be periodic or aperiodic, noise or also superpositions of components. Anyway, what we are measuring is a conglomerate of several components, which means that effects caused by the measuring-devices' electronics and, for example, noise, get added to the signal we are actually after. That is why we have to take the recorded signal, filter out what is of interest to us, and process it. In many cases we are predominantly interested in the periodic components of the signal, or the *spectral content*, which consists of discrete components. For analyses of this kind Fourier transformation is particularly well suited.

Here are some examples:

- Analysis of the vibrations of a violin string or of a bridge,
- checking out the quality of a high-fidelity amplifier,
- radio-frequency Fourier-transformation spectroscopy,
- optical Fourier-transformation spectroscopy,
- digital image-processing (two- and three-dimensional),

to quote only a few examples from acoustics, electronics, and optics, which also shows that this method is not only useful for purely scientific research.

Many mathematical procedures in almost all branches of science and engineering use the Fourier transformation. The method is so widely known—almost "old hat"—that users often only have to push a few buttons (or use a few mouse-clicks) to perform a Fourier transformation, or the lot even gets delivered "to the doorstep, free of charge". This user-friendliness, however, often is accompanied by the loss of all necessary knowledge. Operating errors, incorrect interpretations and frustration result from incorrect settings or similar blunders.

This book aims to raise the level of consciousness concerning the *dos and don'ts* when using Fourier transformation programs. In order to understand how the discrete Fourier transformation programs work it is necessary to discuss Fourier series and continuous Fourier transformations first. Experience shows that mathematical laypersons will have to cope with two hurdles:

- differential and integral calculus and
- complex number arithmetic.

When defining[3] the Fourier series and the continuous Fourier transformation, we cannot help using integrals, as, for example, in Chap. 3 (Window Functions). The problem cannot be avoided, but can be mitigated using integration tables. For example the "Oxford Users' Guide to Mathematics" [6] will be quite helpful in this respect. In Chaps. 4–7 elementary maths will be sufficient to understand what is going on. As far as complex number arithmetic is concerned, I have made sure that in Chap. 1 all formulas are covered in detail, in plain and in complex notation, so this chapter may even serve as a small introduction to dealing with complex numbers.

For all those ready to rip into action using their PCs, the book "Numerical Recipes" [7] is especially useful. It presents, among other things, programs for almost every purpose, and they are commented, too.

---

[3]The definitions given in this book are similar to conventions and do not lay claim to any mathematical rigour.

Chapter 1
Fourier Series

Abstract This chapter introduces the mapping of periodic functions in the time domain to a Fourier series in the frequency domain with Fourier coefficients. It shows how these coefficients are calculated and gives examples. It also provides a gentle introduction to complex notation. It addresses linearity of the transformation, discusses shifting in the time and frequency domain as well as scaling. Parseval's equation is derived. Gibb's phenomena of "ringing" are discussed.

In 1822, Jean Baptiste Joseph Fourier in his "Theorie analytique de la chaleur" described heat transport processes in a circular geometry by a series of sine and cosine functions. Here, we shall first consider much simpler systems, e.g. a string clamped at both ends like in a piano. When excited it will vibrate with its fundamental frequency with the amplitude maximum in the middle while the ends are at rest. This is half a period of a sine-wave. It can also vibrate with twice its fundamental frequency, its so-called first harmonic, with the midpoint at rest, too, and a positive (negative) maximal amplitude at a quarter of the length of the string and a negative (positive) one at three quarters. In fact, it could vibrate with all integer multiples of its fundamental frequency. The possible modes are illustrated in Fig. 4.13 when discussing the sine-transformation. Should you happen to possess a piano, excite a string and feel that the string one octave higher (this is a factor of two) also starts vibrating. It gets excited by the sound (air pressure variations) produced by the first string. The neighbouring strings do not vibrate because their frequencies do not match.

In this chapter we want to be a little more general than a piano string. First, we want to speak of an interval T, corresponding to the length of the string, and continue this interval periodically. Secondly, we want to abandon the constraint that at the interval boundaries the amplitude is zero. Hence, what we are dealing with are arbitrary but periodic functions. Contrary to the clamped string, where we were dealing with half a period and multiples thereof, we now require full periods and their multiples. Thus, when shifting the interval we shall always have a complete period of the function. Let us assume for a moment that the average of the function is zero. Then we may shift the interval boundaries such that the function happens to be zero at the boundaries. A joker might be tempted to put clamps there—and nobody would even notice it! Now it is immediately clear that this arbitrary periodic function can be described by a series of cosines of integer multiples of the fundamental frequency, eventually

© Springer International Publishing Switzerland 2015

T. Butz, *Fourier Transformation for Pedestrians*, Undergraduate Lecture Notes in Physics, DOI 10.1007/978-3-319-16985-9_1

with a phase due to shifting. Instead of a cosine with a phase we can use cosines and sines. Should the average of the function be non-zero, this can be taken care of by a cosine with frequency zero. Finally, we shall consider periodic functions of time (with the exception of Chap. 7), e.g. signals from a measuring device.

Mapping of a *Periodic* Function $f(t)$ to a Series of Fourier Coefficients C_k

1.1 Fourier Series

This section serves as a starter. Many readers may think it too easy; but it should be read and taken seriously all the same. Some preliminary remarks are in order:

i. To make things easier to understand, the whole book will only be concerned with functions in the time domain and their Fourier transforms in the frequency domain. This represents the most common application, and porting it to other pairings, such as space/momentum, for example, is pretty straightforward indeed, as we shall see in Chap. 7.

ii. We use the angular frequency ω when we refer to the frequency domain. The unit of the angular frequency is radians/second (or simpler s^{-1}). It's easily converted to the frequency ν of radio-stations—for example FM 105.4 MHz—using the following equation:

$$\omega = 2\pi\nu. \tag{1.1}$$

The unit of ν is Hz, short for Hertz.

By the way, in case someone wants to do like H.J. Weaver, my much appreciated role-model, and use different notations to avoid having the tedious factors 2π crop up everywhere, don't buy into that. For each 2π you save somewhere, there will be more factors of 2π somewhere else. However, there are valid reasons, as detailed for example in "Numerical Recipes" [7], to use t and ν.

In this book I'll stick to the use of t and ω, cutting down on the cavalier use of 2π that's in vogue elsewhere.

1.1.1 Even and Odd Functions

All functions are either:

$$f(-t) = f(t) : \text{even} \tag{1.2}$$

or:

$$f(-t) = -f(t) : \text{odd} \tag{1.3}$$

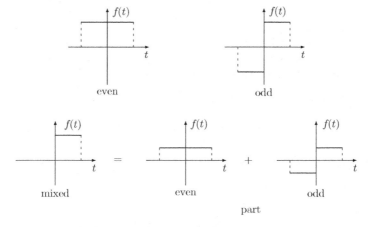

Fig. 1.1 Examples of even, odd and mixed functions

or a "mixture" of both, i.e. even and odd parts superimposed. The decomposition gives:

$$f_{\text{even}}(t) = (f(t) + f(-t))/2$$
$$f_{\text{odd}}(t) = (f(t) - f(-t))/2.$$

See examples in Fig. 1.1.

1.1.2 Definition of the Fourier Series

Fourier analysis is often also called harmonic analysis, as it uses the trigonometric functions sine—an odd function—and cosine—an even function—as basis functions that play a pivotal part in harmonic oscillations.

Similar to expanding a function into a power series, especially periodic functions may be expanded into a series of the trigonometric functions sine and cosine.

Definition 1.1 (*Fourier Series*)

$$f(t) = \sum_{k=0}^{\infty} (A_k \cos \omega_k t + B_k \sin \omega_k t)$$

$$\text{with } \omega_k = \frac{2\pi k}{T} \text{ and } B_0 = 0.$$

(1.4)

Here T means the period of the function $f(t)$. The amplitudes or Fourier coefficients A_k and B_k, are determined in such a way—as we'll see in a moment—that the infinite series is identical with the function $f(t)$. Equation (1.4) therefore tells us, that any periodic function can be represented as a superposition of sine- and cosine-functions with appropriate amplitudes—with an infinite number of terms, if need be—yet using only precisely determined frequencies:

$$\omega = 0, \frac{2\pi}{T}, \frac{4\pi}{T}, \frac{6\pi}{T}, \ldots.$$

Figure 1.2 shows the basis functions for $k = 0, 1, 2, 3$.

Example 1.1 ("Trigonometric identity")

$$f(t) = \cos^2 \omega t = \frac{1}{2} + \frac{1}{2} \cos 2\omega t. \tag{1.5}$$

Trigonometric manipulation in (1.5) already determined the Fourier coefficients A_0 and A_2: $A_0 = 1/2$, $A_2 = 1/2$ (see Fig. 1.3). As function $\cos^2 \omega t$ is an even function, we need no B_k. Generally speaking, practically all "smooth" functions without steps (i.e. without discontinuities) and without kinks (i.e. without discontinuities in their first derivative)—and strictly speaking without discontinuities in all their derivatives—are limited as far as their bandwidth is concerned. This means that a *finite* number of terms in the series will do for practical purposes. Often data gets recorded using a device with limited bandwidth, which puts a limit on how quickly $f(t)$ can vary over time anyway.

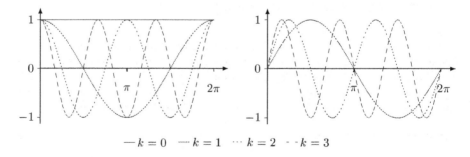

Fig. 1.2 Basis functions of Fourier transformation: cosine (*left*); sine (*right*)

Fig. 1.3 Decomposition of $\cos^2 \omega t$ into the average 1/2 and an oscillation with amplitude 1/2 and frequency 2ω

1.1.3 Calculation of the Fourier Coefficients

Before we dig into the calculation of the Fourier coefficients, we need some tools.

In all following integrals we integrate from $-T/2$ to $+T/2$, meaning over an interval with the period T that is *symmetrical* to $t = 0$. We could also pick any other interval, as long as the integrand is periodic with period T and gets integrated over a *whole* period. The letters n and m in the formulas below are natural numbers $0, 1, 2, \ldots$. Let's have a look at the following:

$$\int_{-T/2}^{+T/2} \cos \frac{2\pi n t}{T} \mathrm{d}t = \begin{cases} 0 & \text{for } n \neq 0 \\ T & \text{for } n = 0 \end{cases}, \tag{1.6}$$

$$\int_{-T/2}^{+T/2} \sin \frac{2\pi n t}{T} \mathrm{d}t = 0 \qquad \text{for all } n. \tag{1.7}$$

This results from the fact, that the areas on the positive half-plane and the ones on the negative one cancel out each other, provided we integrate over a whole number of periods. Cosine integral for $n = 0$ requires special treatment, as it lacks oscillations and therefore areas can't cancel out each other: there the integrand is 1, and the area under the horizontal line is equal to the width of the interval T.

Furthermore, we need the following trigonometric identities:

$$\begin{aligned} \cos \alpha \cos \beta &= 1/2 \ [\cos(\alpha + \beta) + \cos(\alpha - \beta)], \\ \sin \alpha \sin \beta &= 1/2 \ [\cos(\alpha - \beta) - \cos(\alpha + \beta)], \\ \sin \alpha \cos \beta &= 1/2 \ [\sin(\alpha + \beta) + \sin(\alpha - \beta)]. \end{aligned} \tag{1.8}$$

Using these tools we're able to show, without further ado, that the system of basis functions consisting of:

$$1, \ \cos \frac{2\pi t}{T}, \ \sin \frac{2\pi t}{T}, \ \cos \frac{4\pi t}{T}, \ \sin \frac{4\pi t}{T}, \ \ldots, \tag{1.9}$$

is an *orthogonal system.*[1]

Put in formulas, this means:

$$\int_{-T/2}^{+T/2} \cos \frac{2\pi n t}{T} \cos \frac{2\pi m t}{T} dt = \begin{cases} 0 & \text{for } n \neq m \\ T/2 & \text{for } n = m \neq 0, \\ T & \text{for } n = m = 0 \end{cases} \tag{1.10}$$

$$\int_{-T/2}^{+T/2} \sin \frac{2\pi n t}{T} \sin \frac{2\pi m t}{T} dt = \begin{cases} 0 & \text{for } \begin{array}{l} n \neq m, \ n = 0 \\ \text{and/or } m = 0 \end{array}, \\ T/2 & \text{for } n = m \neq 0 \end{cases} \tag{1.11}$$

$$\int_{-T/2}^{+T/2} \cos \frac{2\pi n t}{T} \sin \frac{2\pi m t}{T} dt = 0. \tag{1.12}$$

The right-hand side of (1.10) and (1.11) show that our basis system is not an *orthonormal system,* i.e. the integrals for $n = m$ are not normalised to 1. What's even worse, the special case of (1.10) for $n = m = 0$ is a nuisance, and will keep bugging us again and again.

Using the above orthogonality relations, we're able to calculate the Fourier coefficients straight away. We need to multiply both sides of (1.4) with $\cos \omega_{k'} t$ and integrate from $-T/2$ to $+T/2$. Due to the orthogonality, only terms with $k = k'$ will remain; the second integral will always disappear.

This gives us:

$$A_k = \frac{2}{T} \int_{-T/2}^{+T/2} f(t) \cos \omega_k t \, dt \quad \text{for } k \neq 0 \tag{1.13}$$

and for our "special" case:

$$A_0 = \frac{1}{T} \int_{-T/2}^{+T/2} f(t) dt. \tag{1.14}$$

Please note the prefactors $2/T$ or $1/T$, respectively, in (1.13) and (1.14). Equation (1.14) simply is the average of the function $f(t)$. The "electricians" amongst

---

[1] Similar to two vectors at right angles to each other whose dot product is 0, we call a set of basis functions an orthogonal system if the integral over the product of two different basis functions vanishes.

us, who might think of $f(t)$ as current varying over time, would call A_0 the "DC"-component (DC = direct current, as opposed to AC = alternating current). Now let's multiply both sides of (1.4) with $\sin \omega_{k'}t$ and integrate from $-T/2$ to $+T/2$.

We now have:

$$B_k = \frac{2}{T} \int\limits_{-T/2}^{+T/2} f(t) \sin \omega_k t \, dt \qquad \text{for all } k. \tag{1.15}$$

Equations (1.13) and (1.15) may also be interpreted like so: by weighting the function $f(t)$ with $\cos \omega_{k'}t$ or $\sin \omega_{k'}t$, respectively, we "pick" the spectral components from $f(t)$, when integrating, corresponding to the even or odd components, respectively, of the frequency ω_k. In the following examples we'll only state the functions $f(t)$ in their basic interval $-T/2 \leq t \leq +T/2$. They have to be extended periodically, however, as the definition goes, beyond this basic interval.

Example 1.2 ("Constant") See Fig. 1.4 (*left*):

$$f(t) = 1$$
$$A_0 = 1 \quad \text{"Average"}$$
$$A_k = 0 \quad \text{for all } k \neq 0$$
$$B_k = 0 \quad \text{for all } k \text{ (as } f \text{ is even)}.$$

Example 1.3 ("Triangular function") See Fig. 1.4 (*right*):

$$f(t) = \begin{cases} 1 + \dfrac{2t}{T} & \text{for } -T/2 \leq t \leq 0 \\[2mm] 1 - \dfrac{2t}{T} & \text{for } 0 \leq t \leq +T/2 \end{cases} .$$

Let's recall: $\omega_k = \dfrac{2\pi k}{T}$ $A_0 = 1/2$ ("Average").

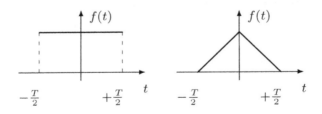

Fig. 1.4 "Constant" (*left*); "triangular function" (*right*). We only show the basic intervals for both functions

For $k \neq 0$ we get:

$$A_k = \frac{2}{T} \left[\int_{-T/2}^{0} \left(1 + \frac{2t}{T}\right) \cos \frac{2\pi kt}{T} dt + \int_{0}^{+T/2} \left(1 - \frac{2t}{T}\right) \cos \frac{2\pi kt}{T} dt \right]$$

$$= \underbrace{\frac{2}{T} \int_{-T/2}^{0} \cos \frac{2\pi kt}{T} dt + \frac{2}{T} \int_{0}^{+T/2} \cos \frac{2\pi kt}{T} dt}_{= 0}$$

$$+ \frac{4}{T^2} \int_{-T/2}^{0} t \cos \frac{2\pi kt}{T} dt - \frac{4}{T^2} \int_{0}^{+T/2} t \cos \frac{2\pi kt}{T} dt$$

$$= -\frac{8}{T^2} \int_{0}^{+T/2} t \cos \frac{2\pi kt}{T} dt.$$

In a last step we'll use $\int x \cos ax \, dx = \frac{x}{a} \sin ax + \frac{1}{a^2} \cos ax$ which finally gives us:

$$A_k = \frac{2(1 - \cos \pi k)}{\pi^2 k^2} \quad (k > 0),$$

$$B_k = 0 \quad \text{(as } f \text{ is even)}.$$

(1.16)

A few more comments on the expression for A_k are in order:

i. For all even k A_k disappears.
ii. For all odd k we get $A_k = 4/(\pi^2 k^2)$.
iii. For $k = 0$ we better use the average A_0 instead of inserting $k = 0$ in (1.16).

We could make things even simpler:

$$A_k = \begin{cases} \dfrac{1}{2} & \text{for } k = 0 \\ \dfrac{4}{\pi^2 k^2} & \text{for k odd} \\ 0 & \text{for k even, } k \neq 0 \end{cases}$$

(1.17)

The series' elements decrease rapidly while k rises (to the power of two in the case of odd k), but in principle we still have an infinite series. That's due to the "pointed roof" at $t = 0$ and the kink (continued periodically!) at $\pm T/2$ in our function $f(t)$. In order to describe these kinks, we need an infinite number of Fourier coefficients.

The following illustrations will show, that things are never as bad as they seem to be:

Using $\omega = 2\pi/T$ (see Fig. 1.5) we get:

$$f(t) = \frac{1}{2} + \frac{4}{\pi^2} \left(\cos \omega t + \frac{1}{9} \cos 3\omega t + \frac{1}{25} \cos 5\omega t \ldots \right). \qquad (1.18)$$

We want to plot the frequencies of this Fourier series. Figure 1.6 shows the result, as produced, for example, by a spectrum analyser,[2] if we would use our "triangular function" $f(t)$ as input signal.

Apart from the DC peak at $\omega = 0$ we can also see the fundamental frequency ω and all odd "harmonics". We may also use this frequency-plot to get an idea about the margins of error resulting from discarding frequencies above, say, 7ω. We will cover this in more detail later on.

1.1.4 Fourier Series in Complex Notation

Let me give you a mild warning before we dig into this chapter: in (1.4) k starts from 0, meaning that we will rule out *negative* frequencies in our Fourier series.

The cosine terms didn't have a problem with negative frequencies. The sign of the cosine argument doesn't matter anyway, so we would be able to go halves, like between brothers, for example, as far as the spectral intensity at the positive frequency $k\omega$ was concerned: $-k\omega$ and $k\omega$ would get equal parts, as shown in Fig. 1.7.

As frequency $\omega = 0$—a frequency as good as any other frequency $\omega \neq 0$—has no "brother", it will not have to go halves. A change of sign for the sine-terms' arguments would result in a change of sign for the corresponding series' term. The splitting of spectral intensity like "between brothers"—equal parts of $-\omega_k$ and $+\omega_k$ now will have to be like "between sisters": the sister for $-\omega_k$ also gets 50%, but her's is *minus* 50%!

Instead of using (1.4) we might as well use:

$$f(t) = \sum_{k=-\infty}^{+\infty} (A_k' \cos \omega_k t + B_k' \sin \omega_k t), \qquad (1.19)$$

where, of course, the following is true: $A_{-k}' = A_k'$, $B_{-k}' = -B_k'$. The formulas for the calculation of A_k' and B_k' for $k > 0$ are identical to (1.13) and (1.15), though they lack the extra factor 2! Equation (1.14) for A_0 stays unaffected by this. This helps us avoid to provide a special treatment for the DC-component.

---

[2]On offer by various companies—for example as a plug-in option for oscilloscopes—for a tidy sum of money.

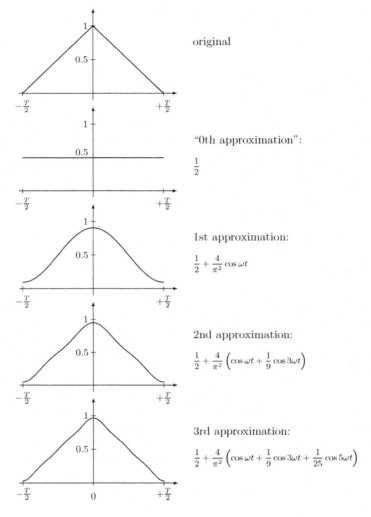

Fig. 1.5 The "triangular function" $f(t)$ and consecutive approximations by a Fourier series with more and more terms

Instead of (1.16) we could have used:

$$A'_k = \frac{(1 - \cos \pi k)}{\pi^2 k^2},\tag{1.20}$$

which would also be valid for $k = 0$! To prove it, we'll use a "dirty trick" or commit a "venial" sin: we'll assume, for the time being, that k is a continuous variable that may steadily decrease towards 0. Then we apply l'Hospital's rule to the expression of

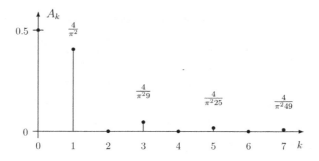

Fig. 1.6 Plot of the "triangular function's" frequencies

Fig. 1.7 Like Fig. 1.6, yet with positive and negative frequencies

type "$0:0$", stating that numerator and denominator may be differentiated separately with respect to k until $\lim_{k\to 0}$ does not result in an expression of type "$0:0$" any more. Like so:

$$\lim_{k\to 0}\frac{1-\cos\pi k}{\pi^2 k^2}=\lim_{k\to 0}\frac{\pi\sin\pi k}{2\pi^2 k}=\lim_{k\to 0}\frac{\pi^2\cos\pi k}{2\pi^2}=\frac{1}{2}. \tag{1.21}$$

If you're no sinner, go for the "average" $A_0 = 1/2$ straight away!

Hint: In many standard Fourier transformation programs a factor 2 between A_0 and $A_{k\neq 0}$ is wrong. This could be mainly due to the fact that frequencies were permitted to be positive only for the basis functions, or positive and negative—like in (1.4). The calculation of the average A_0 is easy as pie, and therefore always recommended as a first test in case of a poorly documented program. As $B_0 = 0$, according to the definition, B_k is a bit harder to check out. Later on we'll deal with simpler checks (for example Parseval's theorem).

Now we're set and ready for the introduction of complex notation. In the following we'll always assume that $f(t)$ is a real function. Generalising this for complex $f(t)$ is no problem. Our most important tool is Euler's identity:

$$e^{i\alpha t}=\cos\alpha t+i\sin\alpha t. \tag{1.22}$$

Here we use i as the imaginary unit that results in -1 when raised to the power of two.

This allows us to rewrite the trigonometric functions as follows:

$$\cos \alpha t = \frac{1}{2}(e^{i\alpha t} + e^{-i\alpha t}),$$
$$\sin \alpha t = \frac{1}{2i}(e^{i\alpha t} - e^{-i\alpha t}). \tag{1.23}$$

Inserting into (1.4) gives:

$$f(t) = A_0 + \sum_{k=1}^{\infty} \left(\frac{A_k - iB_k}{2} e^{i\omega_k t} + \frac{A_k + iB_k}{2} e^{-i\omega_k t} \right). \tag{1.24}$$

Using the short-cuts:

$$C_0 = A_0,$$
$$C_k = \frac{A_k - iB_k}{2},$$
$$C_{-k} = \frac{A_k + iB_k}{2}, \qquad k = 1, 2, 3, \ldots, \tag{1.25}$$

we finally get:

$$f(t) = \sum_{k=-\infty}^{+\infty} C_k e^{i\omega_k t}, \qquad \omega_k = \frac{2\pi k}{T}. \tag{1.26}$$

Caution: For $k < 0$ there will be *negative* frequencies. (No worries, according to our above digression!) Pretty handy that C_k and C_{-k} are complex conjugates to each other (see "brother and sister".) Now C_k can be formulated just as easily:

$$C_k = \frac{1}{T} \int_{-T/2}^{+T/2} f(t) e^{-i\omega_k t} dt \qquad \text{for } k = 0, \pm 1, \pm 2, \ldots. \tag{1.27}$$

Please note that there is a negative sign in the exponent. It will stay with us till the end of this book, and, hopefully, for the rest of your life. Please also note that the index k runs from $-\infty$ to $+\infty$ for C_k whereas it runs from 0 to $+\infty$ for A_k and B_k.

1.2 Theorems and Rules

1.2.1 Linearity Theorem

Expanding a periodic function into a Fourier series is a linear operation. This means, that we may use the two Fourier pairs:

$$f(t) \leftrightarrow \{C_k; \omega_k\} \text{ and}$$
$$g(t) \leftrightarrow \{C_k'; \omega_k\}$$

to form the following linear combination:

$$h(t) = af(t) + bg(t) \leftrightarrow \{aC_k + bC_k'; \omega_k\}. \tag{1.28}$$

Thus we may easily determine the Fourier series of a function by splitting it into items whose Fourier series we already know.

Example 1.4 (Lowered "triangular function") The simplest example is our "triangular function" from Example 1.3, though this time it is symmetrical to its base line (see Fig. 1.8): we only have to subtract $1/2$ from our original function. That means, that the Fourier series remained unchanged while only the average A_0 now turned to 0.

The linearity theorem appears to be so trivial that you may accept it at face-value even when you have "strayed from the path of virtue". Straying from the path of virtue is, for example, something as elementary as squaring.

1.2.2 The First Shifting Rule (Shifting Within the Time Domain)

Often we want to know how the Fourier series changes if we shift the function $f(t)$ along the time axis. This, for example, happens on a regular basis if we use a different interval, e.g. from 0 to T, instead of the symmetrical one from $-T/2$ to $T/2$ we have used so far. In this situation, the First Shifting Rule comes in very handy:

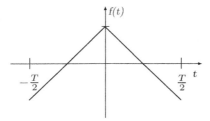

Fig. 1.8 "Triangular function" with average 0

$$f(t) \leftrightarrow \{C_k; \omega_k\},$$
$$f(t - a) \leftrightarrow \{C_k e^{-i\omega_k a}; \omega_k\}. \tag{1.29}$$

Proof (*First Shifting Rule*)

$$C_k^{\text{new}} = \frac{1}{T} \int\limits_{-T/2}^{+T/2} f(t - a)e^{-i\omega_k t} dt = \frac{1}{T} \int\limits_{-T/2-a}^{+T/2-a} f(t')e^{-i\omega_k t'}e^{-i\omega_k a} dt'$$

$$= e^{-i\omega_k a} C_k^{\text{old}}. \quad \square$$

We integrate over a full period, that's why shifting the limits of the interval by a does not make any difference.

The proof is trivial, the result of the shifting along the time axis not! The new Fourier coefficient results from the old coefficient C_k by multiplying it with the phase factor $e^{-i\omega_k a}$. As C_k generally is complex, shifting "shuffles" real and imaginary parts.

Without using complex notation we get:

$$f(t) \leftrightarrow \{A_k; B_k; \omega_k\},$$
$$f(t - a) \leftrightarrow \{A_k \cos \omega_k a - B_k \sin \omega_k a; A_k \sin \omega_k a + B_k \cos \omega_k a; \omega_k\}. \tag{1.30}$$

Two examples follow:

Example 1.5 (Quarter period shifted "triangular function") "Triangular function" (with average = 0) (see Fig. 1.8):

$$f(t) = \begin{cases} \dfrac{1}{2} + \dfrac{2t}{T} & \text{for } -T/2 \le t \le 0 \\[2mm] \dfrac{1}{2} - \dfrac{2t}{T} & \text{for } 0 < t \le T/2 \end{cases}$$

$$\text{with } C_k = \begin{cases} \dfrac{1 - \cos \pi k}{\pi^2 k^2} = \dfrac{2}{\pi^2 k^2} & \text{for } k \text{ odd} \\[2mm] 0 & \text{for } k \text{ even} \end{cases}. \tag{1.31}$$

Now let's shift this function to the right by $a = T/4$:

$$f_{\text{new}} = f_{\text{old}}(t - T/4).$$

So the new coefficients can be calculated as follows:

$$
\begin{aligned}
C_k^{\text{new}} &= C_k^{\text{old}} e^{-i\pi k/2} && (k \text{ odd}) \\
&= \frac{2}{\pi^2 k^2} \left(\cos \frac{\pi k}{2} - i \sin \frac{\pi k}{2} \right) && (k \text{ odd}) \\
&= -\frac{2i}{\pi^2 k^2} (-1)^{\frac{k-1}{2}} && (k \text{ odd}).
\end{aligned}
\tag{1.32}
$$

It's easy to realise that $C_{-k}^{\text{new}} = -C_k^{\text{new}}$.
In other words: $A_k = 0$.
Using $iB_k = C_{-k} - C_k$ we finally get:

$$
B_k^{\text{new}} = \frac{4}{\pi^2 k^2} (-1)^{\frac{k+1}{2}} \qquad (k \text{ odd}).
$$

Using the above shifting we get an odd function (see Fig. 1.9b).

Example 1.6 (Half period shifted "triangular function") Now we'll shift the same function to the right by $a = T/2$:

$$
f_{\text{new}} = f_{\text{old}}(t - T/2).
$$

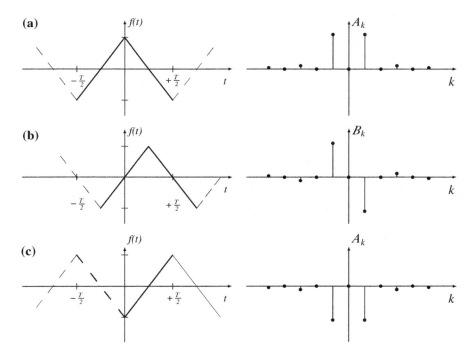

Fig. 1.9 **a** "Triangular function" (with average $= 0$); **b** right-shifted by $T/4$; **c** right-shifted by $T/2$

The new coefficients then are:

$$
\begin{aligned}
C_k^{\text{new}} &= C_k^{\text{old}} e^{-i\pi k} && (k \;\; \text{odd})\\
&= \frac{2}{\pi^2 k^2}\,(\cos \pi k - i \sin \pi k) && (k \;\; \text{odd})\\
&= -\frac{2}{\pi^2 k^2} && (k \;\; \text{odd})
\end{aligned}
\tag{1.33}
$$

$$(C_0 = 0 \;\; \text{stays}).$$

So we've only changed the sign. That's okay, as the function now is upside-down (see Fig. 1.9c).

Warning: Shifting by $a = T/4$ will result in alternating signs for the coefficients (Fig. 1.9b). The series of Fourier coefficients, that are decreasing monotonically with k according to Fig. 1.9a, looks pretty "frazzled" after shifting the function by $a = T/4$, due to the alternating sign.

1.2.3 The Second Shifting Rule (Shifting Within the Frequency Domain)

The First Shifting Rule showed us that shifting within the time domain leads to a multiplication by a phase factor in the frequency domain. Reversing this statement gives us the Second Shifting Rule:

$$
\begin{aligned}
f(t) &\leftrightarrow \{C_k; \omega_k\},\\
f(t)e^{i\frac{2\pi a t}{T}} &\leftrightarrow \{C_{k-a}; \omega_k\}.
\end{aligned}
\tag{1.34}
$$

In other words: a multiplication of the function $f(t)$ by the phase factor $e^{i2\pi a t/T}$ results in frequency ω_k now being related to "shifted" coefficient C_{k-a}—instead of the former coefficient C_k. A comparison between (1.34) and (1.29) demonstrates the two-sided character of the two Shifting Rules. If a is an integer, there won't be any problem if you simply take the coefficient shifted by a. But what if a is not an integer?

Strangely enough nothing serious will happen. Simply shifting like we did before won't work any more, but who is to keep us from inserting $(k-a)$ into the expression for old C_k, whenever k occurs.

(If it's any help to you, do commit another venial sin and temporarily consider k to be a continuous variable.) So in the case of non-integer a we didn't really shift C_k, but rather recalculated it using shifted k.

Caution: If you have simplified a k-dependency in the expressions for C_k, for example:

$$1 - \cos \pi k = \begin{cases} 0 & \text{for k even} \\ 2 & \text{for k odd} \end{cases}$$

(as in (1.16)), you'll have trouble replacing the "vanished" k with $(k-a)$. In this case there's only one way out: back to the expressions with *all* k-dependencies *without* simplification.

Before we present examples, two more ways of writing down the Second Shifting Rule are in order:

$$f(t) \leftrightarrow \{A_k; B_k; \omega_k\},$$

$$f(t)e^{\frac{2\pi iat}{T}} \leftrightarrow \left\{ \frac{1}{2}[A_{k+a} + A_{k-a} + i(B_{k+a} - B_{k-a})]; \right. \tag{1.35}$$

$$\left. \frac{1}{2}[B_{k+a} + B_{k-a} + i(A_{k-a} - A_{k+a})]; \omega_k \right\}.$$

Caution: This is true for $k \neq 0$.

Old A_0 then becomes $A_a/2 + iB_a/2$!

This is easily proved by solving (1.25) for A_k and B_k and inserting it in (1.34):

$$\begin{aligned} A_k &= C_k + C_{-k}, \\ -iB_k &= C_k - C_{-k}, \end{aligned} \tag{1.36}$$

$$A_k^{\text{new}} = C_k + C_{-k} = \frac{A_{k-a} - iB_{k-a}}{2} + \frac{A_{k+a} + iB_{k+a}}{2},$$

$$-iB_k^{\text{new}} = C_k - C_{-k} = \frac{A_{k-a} - iB_{k-a}}{2} - \frac{A_{k+a} + iB_{k+a}}{2},$$

which leads to (1.35). We get the special treatment for A_0 from:

$$A_0^{\text{new}} = C_0^{\text{new}} = \frac{A_{-a} - iB_{-a}}{2} = \frac{A_{+a} + iB_{+a}}{2}.$$

The formulas become a lot simpler in case $f(t)$ is real. Then we get:

$$f(t) \cos \frac{2\pi at}{T} \leftrightarrow \left\{ \frac{A_{k+a} + A_{k-a}}{2}; \frac{B_{k+a} + B_{k-a}}{2}; \omega_k \right\}, \tag{1.37}$$

old A_0 becomes $A_a/2$ and also:

$$f(t) \sin \frac{2\pi a t}{T} \leftrightarrow \left\{ \frac{B_{k+a} - B_{k-a}}{2}; \frac{A_{k-a} - A_{k+a}}{2}; \omega_k \right\},$$

old A_0 becomes $B_a/2$.

Example 1.7 ("Constant")

$$f(t) = 1 \quad \text{for} \ -T/2 \le t \le +T/2.$$

$A_k = \delta_{k,0}$ (Kronecker symbol, see Sect. 4.1.2) or $A_0 = 1$, all other A_k, B_k vanish. Of course we've always known that $f(t)$ is a cosine wave with frequency $\omega = 0$ and therefore only required the coefficient for $\omega = 0$.

Now let's multiply function $f(t)$ by $\cos(2\pi t/T)$, i.e. $a = 1$. From (1.37) we can see:

$$A_k^{\text{new}} = \delta_{k-1,0}, \quad \text{i.e.} \quad A_1 = 1 \ (\text{all others are } 0),$$
$$\text{or} \quad C_1 = 1/2, \quad\quad\quad C_{-1} = 1/2.$$

So we have shifted the coefficient by $a = 1$ (to the right and to the left, and gone halves, like "between brothers").

This example demonstrates that the frequency $\omega = 0$ is as good as any other frequency. No kidding! If you know, for example, the Fourier series of a function $f(t)$ and consequently the solution for integrals of the form:

$$\int\limits_{-T/2}^{+T/2} f(t) e^{-i\omega_k t} dt$$

then you already have, using the Second Shifting Rule, solved all integrals for $f(t)$, multiplied by $\sin(2\pi a t/T)$ or $\cos(2\pi a t/T)$. No wonder, you only had to combine phase factor $e^{i2\pi a t/T}$ with phase factor $e^{-i\omega_k t}$!

Example 1.8 ("Triangular function" multiplied by cosine) The function:

$$f(t) = \begin{cases} 1 + \dfrac{2t}{T} & \text{for} \ -T/2 \le t \le 0 \\[2mm] 1 - \dfrac{2t}{T} & \text{for} \ 0 \le t \le T/2 \end{cases}$$

is to be multiplied by $\cos(\pi t/T)$, i.e. we shift the coefficients C_k by $a = 1/2$ (see Fig. 1.10). The new function still is even, and therefore we only have to look after A_k:

$$A_k^{\text{new}} = \frac{A_{k+a}^{\text{old}} + A_{k-a}^{\text{old}}}{2}.$$

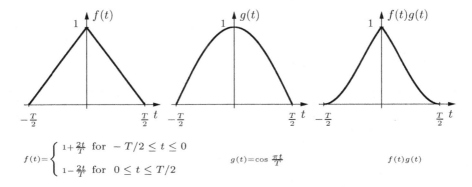

$$f(t)=\begin{cases} 1+\frac{2t}{T} & \text{for } -T/2 \le t \le 0 \\ 1-\frac{2t}{T} & \text{for } 0 \le t \le T/2 \end{cases}$$

$$g(t)=\cos\frac{\pi t}{T}$$

$$f(t)g(t)$$

Fig. 1.10 "Triangular function" (*left*); $\left(\cos\frac{\pi t}{T}\right)$-function (*middle*); "triangular function" with $\left(\cos\frac{\pi t}{T}\right)$-weighting (*right*)

We use (1.16) for the old A_k (and stop using the simplified version (1.17)!):

$$A_k^{\text{old}} = \frac{2(1-\cos\pi k)}{\pi^2 k^2}.$$

We then get:

$$\begin{aligned} A_k^{\text{new}} &= \frac{1}{2}\left[\frac{2(1-\cos\pi(k+1/2))}{\pi^2(k+1/2)^2} + \frac{2(1-\cos\pi(k-1/2))}{\pi^2(k-1/2)^2}\right] \\ &= \frac{1-\cos\pi k\cos(\pi/2)+\sin\pi k\sin(\pi/2)}{\pi^2(k+1/2)^2} \\ &\quad + \frac{1-\cos\pi k\cos(\pi/2)-\sin\pi k\sin(\pi/2)}{\pi^2(k-1/2)^2} \\ &= \frac{1}{\pi^2(k+1/2)^2} + \frac{1}{\pi^2(k-1/2)^2} \end{aligned}$$

$$(1.38)$$

$$A_0^{\text{new}} = \frac{A_{1/2}^{\text{old}}}{2} = \frac{2(1-\cos(\pi/2))}{2\pi^2\left(\frac{1}{2}\right)^2} = \frac{4}{\pi^2}.$$

The new coefficients then are:

$$A_0 = \frac{4}{\pi^2},$$

$$A_1 = \frac{1}{\pi^2}\left(\frac{1}{\left(\frac{3}{2}\right)^2} + \frac{1}{\left(\frac{1}{2}\right)^2}\right) = \frac{4}{\pi^2}\left(\frac{1}{9} + \frac{1}{1}\right) = \frac{4}{\pi^2}\frac{10}{9},$$

$$A_2 = \frac{1}{\pi^2}\left(\frac{1}{\left(\frac{5}{2}\right)^2} + \frac{1}{\left(\frac{3}{2}\right)^2}\right) = \frac{4}{\pi^2}\left(\frac{1}{25} + \frac{1}{9}\right) = \frac{4}{\pi^2}\frac{34}{225}, \qquad (1.39)$$

$$A_3 = \frac{1}{\pi^2}\left(\frac{1}{\left(\frac{7}{2}\right)^2} + \frac{1}{\left(\frac{5}{2}\right)^2}\right) = \frac{4}{\pi^2}\left(\frac{1}{49} + \frac{1}{25}\right) = \frac{4}{\pi^2}\frac{74}{1225} \text{ etc.}$$

A comparison of these coefficients with the ones without the $\left(\cos\frac{\pi t}{T}\right)$-weighting shows what we've done:

	without weighting	with $\left(\cos\frac{\pi t}{T}\right)$-weighting
A_0	$\frac{1}{2}$	$\frac{4}{\pi^2}$
A_1	$\frac{4}{\pi^2}$	$\frac{4}{\pi^2}\frac{10}{9}$
A_2	0	$\frac{4}{\pi^2}\frac{34}{225}$
A_3	$\frac{4}{\pi^2}\frac{1}{9}$	$\frac{4}{\pi^2}\frac{74}{1225}.$

$$(1.40)$$

We can see the following:

i. The average A_0 got somewhat smaller, as the rising and falling flanks were weighted with the cosine, which, except for $t = 0$, is less than 1.
ii. We raised coefficient A_1 a bit, but lowered all following odd coefficients a bit, too. This is evident straight away, if we convert:

$$\frac{1}{(2k+1)^2} + \frac{1}{(2k-1)^2} < \frac{1}{k^2} \quad \text{to} \quad 8k^4 - 10k^2 + 1 > 0.$$

This is not valid for $k = 1$, yet all bigger k.
iii. Now we've been landed with even coefficients, that were 0 before.

We now have twice as many terms in the series as before, though they go down at an increased rate when k increases. The multiplication by $\cos(\pi t/T)$ caused the kink at $t = 0$ to turn into a much more pointed spike. This should actually make for a worsening of convergence or a slower rate of decrease of the coefficients. We have, however, rounded the kink at the interval-boundary $\pm T/2$, which naturally helps, but we couldn't reasonably have predicted what exactly was going to happen.

1.2.4 Scaling Theorem

Sometimes we happen to want to scale the time axis. In this case, there is no need to re-calculate the Fourier coefficients. From:

$$f(t) \leftrightarrow \{C_k; \omega_k\}$$
$$f(at) \leftrightarrow \{C_k; a \cdot \omega_k\}. \tag{1.41}$$

Here, a must be real!

For $a < 1$ the time axis will be stretched and, hence, the frequency axis will be compressed. For $a > 1$ the opposite is true. The proof for (1.41) is easy and follows from (1.27). Please note that we also have to stretch or compress the interval limits because of the requirement of periodicity. Similarly, the basis function are modified according to $\omega_k^{\text{new}} = a \cdot \omega_k^{\text{old}}$.

$$C_k^{\text{new}} = \frac{a}{T} \int\limits_{-T/2a}^{+T/2a} f(at) e^{-i\omega_k^{\text{new}} t}\, dt = \frac{a}{T} \int\limits_{-T/2}^{+T/2} f(t') e^{-i\omega_k^{\text{old}} t'} \frac{1}{a}\, dt' = C_k^{\text{old}}.$$

$$\text{with } t' = at$$
$$\text{and } \omega_k^{\text{new}} t = \omega_k^{\text{old}} t'$$

Here, we have tacitly assumed $a > 0$. For $a < 0$, we would only reverse the time axis and, hence, also the frequency axis. For the special case $a = -1$ we have:

$$f(t) \leftrightarrow \{C_k, \omega_k\},$$
$$f(-t) \leftrightarrow \{C_k; -\omega_k\}. \tag{1.42}$$

1.3 Partial Sums, Bessel's Inequality, Parseval's Equation

For practical work, infinite Fourier series have to get terminated at some stage, regardless. Therefore we only use a partial sum, say until we reach $k_{\text{max}} = N$. This Nth partial sum then is:

$$S_N = \sum_{k=0}^{N} (A_k \cos \omega_k t + B_k \sin \omega_k t). \tag{1.43}$$

Terminating the series results in the following squared error:

$$\delta_N^2 = \frac{1}{T} \int\limits_T [f(t) - S_N(t)]^2 dt. \tag{1.44}$$

The "T" below the integral symbol means integration over a full period. This definition will become plausible in a second if we look at the discrete version:

$$\delta_N^2 = \frac{1}{N} \sum_{i=1}^{N} (f_i - s_i)^2.$$

Please note that we divide by the length of the interval, to compensate for integrating over the interval T. Now we know that the following is correct for the infinite series:

$$\lim_{N \to \infty} S_N = \sum_{k=0}^{\infty} (A_k \cos \omega_k t + B_k \sin \omega_k t) \qquad (1.45)$$

provided the A_k and B_k happen to be the Fourier coefficients. Does this also have to be true for the Nth partial sum? Isn't there a chance the mean squared error would get smaller, if we used other coefficients instead of Fourier coefficients? That's not the case! To prove it, we'll now insert (1.43) and (1.44) in (1.45), leave out $\lim_{N \to \infty}$ and get:

$$
\begin{aligned}
\delta_N^2 &= \frac{1}{T} \left\{ \int_T f^2(t)\mathrm{d}t - 2 \int_T f(t)S_N(t)\mathrm{d}t + \int_T S_N^2(t)\mathrm{d}t \right\} \\
&= \frac{1}{T} \left\{ \int_T f^2(t)\mathrm{d}t \right. \\
&\quad - 2 \int_T \sum_{k=0}^{\infty} (A_k \cos \omega_k t + B_k \sin \omega_k t) \sum_{k=0}^{N} (A_k \cos \omega_k t + B_k \sin \omega_k t)\mathrm{d}t \\
&\quad \left. + \int_T \sum_{k=0}^{N} (A_k \cos \omega_k t + B_k \sin \omega_k t) \sum_{k=0}^{N} (A_k' \cos \omega_k' t + B_k' \sin \omega_k' t)\mathrm{d}t \right\} \\
&= \frac{1}{T} \left\{ \int_T f^2(t)\mathrm{d}t - 2T A_0^2 - 2\frac{T}{2} \sum_{k=1}^{N} (A_k^2 + B_k^2) + T A_0^2 \right. \\
&\quad \left. + \frac{T}{2} \sum_{k=1}^{N} (A_k^2 + B_k^2) \right\} \\
&= \frac{1}{T} \int_T f^2(t)\mathrm{d}t - A_0^2 - \frac{1}{2} \sum_{k=1}^{N} (A_k^2 + B_k^2). \qquad (1.46)
\end{aligned}
$$

Here we made use of the somewhat cumbersome orthogonality properties of (1.10)–(1.12). As the A_k^2 and B_k^2 always are positive, the mean squared error will drop *monotonically* as N increases.

Example 1.9 (Approximating the "triangular function") The "Triangular function":

$$f(t) = \begin{cases} 1 + \dfrac{2t}{T} & \text{for } -T/2 \le t \le 0 \\ 1 - \dfrac{2t}{T} & \text{for } 0 \le t \le T/2 \end{cases} \tag{1.47}$$

has the mean squared "signal":

$$\frac{1}{T} \int_{-T/2}^{+T/2} f^2(t)dt = \frac{2}{T} \int_{0}^{+T/2} f^2(t)dt = \frac{2}{T} \int_{0}^{+T/2} \left(1 - 2\frac{t}{T}\right)^2 dt = \frac{1}{3}. \tag{1.48}$$

The most coarse, meaning 0th, approximation is:

$$S_0 = 1/2, \text{ i.e.}$$
$$\delta_0^2 = 1/3 - 1/4 = 1/12 = 0.0833 \ldots.$$

The next approximation results in:

$$S_1 = 1/2 + \tfrac{4}{\pi^2} \cos \omega t, \text{ i.e.}$$
$$\delta_1^2 = 1/3 - 1/4 - 1/2 \left(\tfrac{4}{\pi^2}\right)^2 = 0.0012 \ldots.$$

For δ_3^2 we get $0.0001915\ldots$, the approximation of the partial sum to the "triangle" quickly gets better and better.

As δ_N^2 is always positive, we finally arrive from (1.46) at Bessel's inequality:

$$\frac{1}{T} \int_T f^2(t)dt \ge A_0^2 + \frac{1}{2}\sum_{k=1}^{N}(A_k^2 + B_k^2). \tag{1.49}$$

For the limiting case of $N \to \infty$ we get Parseval's equation:

$$\frac{1}{T} \int_T f^2(t)dt = A_0^2 + \frac{1}{2}\sum_{k=1}^{\infty}(A_k^2 + B_k^2). \tag{1.50}$$

Parseval's equation may be interpreted as follows: $1/T \int f^2(t)dt$ is the mean squared "signal" within the time domain, or—more colloquially—the "information content". Fourier series don't lose this information content: it's in the squared Fourier coefficients.

The rule of thumb therefore is:

"The information content isn't lost."
or
"Nothing goes missing in this house."

Here we simply have to mention an analogy with the energy density of the electromagnetic field: $w = \frac{1}{2}(E^2 + B^2)$ with $\epsilon_0 = \mu_0 = 1$, as often is customary in theoretical physics. The comparison has got some weak sides, as E and B have nothing to do with even and odd components.

Parseval's equation is very useful: you can use it to easily sum up infinite series. I think you'd always have been curious how we arrive at formulas such as, for example,

$$\sum_{\substack{k=1 \\ \text{odd}}}^{\infty} \frac{1}{k^4} = \frac{\pi^4}{96}. \tag{1.51}$$

Our "triangular function" (1.47) is behind it! Insert (1.48) and (1.17) in (1.50), and you'll get:

$$\frac{1}{3} = \frac{1}{4} + \frac{1}{2} \sum_{\substack{k=1 \\ \text{odd}}}^{\infty} \left(\frac{4}{\pi^2 k^2}\right)^2 \tag{1.52}$$

$$\text{or} \sum_{\substack{k=1 \\ \text{odd}}}^{\infty} \frac{1}{k^4} = \frac{2}{12} \cdot \frac{\pi^4}{16} = \frac{\pi^4}{96}.$$

1.4 Gibbs' Phenomenon

So far we've only been using smooth functions as examples for $f(t)$, or—like the much-used "triangular function"—functions with "a kink", that's a discontinuity in the first derivative. This pointed kink made sure that we basically needed an infinite number of terms in the Fourier series. Now, what will happen if there is a step, a discontinuity, in the function itself? This certainly won't make the problem with the infinite number of elements any smaller. Is there any way to approximate such a step by using the Nth partial sum, and will the mean squared error for $N \to \infty$ approach 0? The answer is clearly **"Yes and No"**. Yes, because it apparently works, and no, because Gibbs' phenomenon happens at the steps, an overshoot or undershoot, that doesn't disappear for $N \to \infty$.

In order to understand this, we'll have to dig a bit wider.

1.4.1 Dirichlet's Integral Kernel

The following expression is called Dirichlet's integral kernel:

$$D_N(x) = \frac{\sin\left(N + \frac{1}{2}\right) x}{2 \sin \frac{x}{2}}$$

$$= \tfrac{1}{2} + \cos x + \cos 2x + \cdots + \cos Nx. \tag{1.53}$$

The second equal sign can be proved as follows:

$$\begin{aligned}
\left(2 \sin \tfrac{x}{2}\right) D_N(x) &= 2 \sin \tfrac{x}{2} \times \left(\tfrac{1}{2} + \cos x + \cos 2x + \cdots + \cos Nx\right) \\
&= \sin \tfrac{x}{2} + 2 \cos x \sin \tfrac{x}{2} + 2 \cos 2x \sin \tfrac{x}{2} + \cdots \\
&\quad + 2 \cos Nx \sin \tfrac{x}{2} \\
&= \sin \left(N + \tfrac{1}{2}\right) x.
\end{aligned} \tag{1.54}$$

Here we have used the identity:

$$2 \sin \alpha \cos \beta = \sin(\alpha + \beta) + \sin(\alpha - \beta)$$
$$\text{with } \alpha = x/2 \text{ and } \beta = nx, \qquad n = 1, 2, \dots, N.$$

By insertion, we see that all pairs of terms cancel out each other, except for the last one.

Figure 1.11 shows a few examples for $D_N(x)$. Please note that $D_N(x)$ is periodic in 2π. This is immediately evident from the cosine notation. With $x = 0$ we get $D_N(0) = N + 1/2$, between 0 and 2π $D_N(x)$ oscillates around 0.

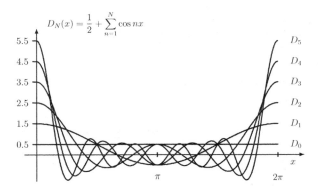

Fig. 1.11 $D_N(x) = 1/2 + \cos x + \cos 2x + \cdots + \cos Nx$

In the limiting case of $N \to \infty$ everything averages to 0, except for $x = 0$ (modulo 2π), that's where $D_N(x)$ grows beyond measure. Here we've found a notation for the δ-function (see Chap. 2)! Please excuse even two venial sins I've committed here: first, the δ-function is a distribution (and not a function!), and second, $\lim_{N \to \infty} D_N(x)$ is a whole "comb" of δ-functions 2π apart.

1.4.2 Integral Notation of Partial Sums

We need a way to sneak up on the discontinuity, from the left and the right. That's why we insert the defining equations for the Fourier coefficients, (1.13)–(1.15), in (1.43):

$$S_N(t) = \frac{1}{T} \int_{-T/2}^{+T/2} f(x)\mathrm{d}x \quad \left\{ \begin{array}{l} (k = 0)\text{-term taken out} \\ \text{of the sum} \end{array} \right.$$

$$+ \sum_{k=1}^{N} \frac{2}{T} \int_{-T/2}^{+T/2} \left(f(x) \cos \frac{2\pi kx}{T} \cos \frac{2\pi kt}{T} \right.$$

$$\left. + f(x) \sin \frac{2\pi kx}{T} \sin \frac{2\pi kt}{T} \right) \mathrm{d}x \qquad (1.55)$$

$$= \frac{2}{T} \int_{-T/2}^{+T/2} f(x) \left(\frac{1}{2} + \sum_{k=1}^{N} \cos \frac{2\pi k(x - t)}{T} \right) \mathrm{d}x$$

$$= \frac{2}{T} \int_{-T/2}^{+T/2} f(x) D_N \left(\frac{2\pi(x-t)}{T} \right) \mathrm{d}x.$$

Using the substitution $x - t = u$ we get:

$$S_N(t) = \frac{2}{T} \int_{-T/2-t}^{+T/2-t} f(u + t) D_N \left(\frac{2\pi u}{T} \right) \mathrm{d}u. \qquad (1.56)$$

As both f and D are periodic in T, we may shift the integration boundaries by t with impunity, without changing the integral. Now we split the integration interval from $-T/2$ to $+T/2$:

$$S_N(t) = \frac{2}{T} \left\{ \int\limits_{-T/2}^{0} f(u+t)D_N(\tfrac{2\pi u}{T})\mathrm{d}u + \int\limits_{0}^{+T/2} f(u+t)D_N(\tfrac{2\pi u}{T})\mathrm{d}u \right\}$$

$$= \frac{2}{T} \int\limits_{0}^{+T/2} [f(t-u) + f(t+u)]D_N(\tfrac{2\pi u}{T})\mathrm{d}u. \tag{1.57}$$

Here we made good use of the fact that D_N is an even function (sum over cosine terms!).

Riemann's localisation theorem—which we won't prove here in the scientific sense, but which can be understood straight away using (1.57)—states that the convergence behaviour of $S_N(t)$ for $N \to \infty$ only depends on the immediate proximity to t of the function:

$$\lim_{N\to\infty} S_N(t) = S(t) = \frac{f(t^+) + f(t^-)}{2}. \tag{1.58}$$

Here t^+ and t^- mean the approach to t, from above and from below. Contrary to a continuous function with a non-differentiability ("kink"), where $\lim_{N\to\infty} S_N(t) = f(t)$, (1.58) means, that in the case of a discontinuity ("step") at t, the partial sum converges to a value that's "half-way" there.

That seems to make sense.

1.4.3 Gibbs' Overshoot

Now we'll have a closer look at the unit step (see Fig. 1.12):

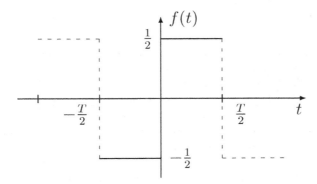

Fig. 1.12 Unit step

$$f(t) = \begin{cases} -1/2 & \text{for } -T/2 \le t < 0 \\ +1/2 & \text{for } 0 \le t \le T/2 \end{cases} \text{ with periodic continuation.} \qquad (1.59)$$

At this stage we're only interested in the case where $t > 0$, and $t \le T/4$. The factor in (1.57) preceding to Dirichlet's integral kernel is:

$$f(t-u) + f(t+u) = \begin{cases} 1 & \text{for } 0 \le u < t \\ 0 & \text{for } t \le u < T/2 - t \\ -1 & \text{for } T/2 - t \le u < T/2 \end{cases} . \qquad (1.60)$$

Inserting in (1.57) results in:

$$\begin{aligned}
S_N(t) &= \frac{2}{T} \left\{ \int_0^t D_N(\tfrac{2\pi u}{T}) du \;-\; \int_{T/2-t}^{T/2} D_N(\tfrac{2\pi u}{T}) du \right\} \\
&= \left\{ \frac{1}{\pi} \int_0^{2\pi t/T} D_N(x) dx \;-\; \int_{-2\pi t/T}^0 D_N(x - \pi) dx \right\} \\
&\quad\;\; (\text{with } x = \tfrac{2\pi u}{T}) \qquad\quad (\text{with } x = \tfrac{2\pi u}{T} - \pi).
\end{aligned} \qquad (1.61)$$

Now we'll insert the expression of Dirichlet's kernel as sum of cosine terms and integrate them:

$$\begin{aligned}
S_N(t) &= \frac{1}{\pi} \left\{ \frac{\pi t}{T} + \frac{\sin \frac{2\pi t}{T}}{1} + \frac{\sin 2\frac{2\pi t}{T}}{2} + \cdots + \frac{\sin N \frac{2\pi t}{T}}{N} \right. \\
&\qquad \left. - \left(\frac{\pi t}{T} - \frac{\sin \frac{2\pi t}{T}}{1} + \frac{\sin 2\frac{2\pi t}{T}}{2} - \cdots + (-1)^N \frac{\sin N \frac{2\pi t}{T}}{N} \right) \right\} \\
&= \frac{2}{\pi} \sum_{\substack{k=1 \\ \text{odd}}}^{N} \frac{1}{k} \sin \frac{2\pi k t}{T}.
\end{aligned} \qquad (1.62)$$

This function is the expression of the partial sums of the unit step. In Fig. 1.13 we show some approximations.

Figure 1.14 shows the 49th partial sum. As we can see, we're already getting pretty close to the unit step, but there are overshoots and undershoots near the discontinuity. Electro-technical engineers know this phenomenon when using filters with very steep flanks: the signal "rings". We could be led to believe that the amplitude of these overshoots and undershoots will get smaller and smaller, provided only we make N big enough. We haven't got a chance! Comparing Fig. 1.13 with Fig. 1.14 should have made us think twice. We'll have a closer look at that, using the following approximation: N is to be very big and t (or x in (1.61), respectively) very small, i.e. close to 0.

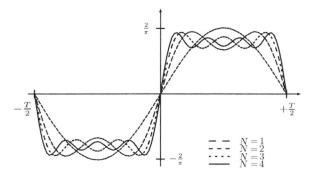

Fig. 1.13 Partial sum expression of unit step

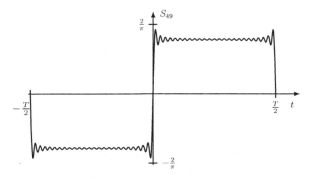

Fig. 1.14 Partial sum expression of unit step for $N = 49$

Then we may neglect $1/2$ with respect to N in the numerator of Dirichlet's kernel and replace $\sin(x/2)$ by $x/2$:

$$D_N(x) \rightarrow \frac{\sin Nx}{x}. \tag{1.63}$$

Therefore the partial sum for large N and close to $t = 0$ becomes:

$$S_N(t) \rightarrow \frac{1}{\pi} \int\limits_0^{2\pi Nt/T} \frac{\sin z}{z} dz \tag{1.64}$$

with $z = Nx$.

That is the sine integral. We'll get the extremes at $\dfrac{dS_N(t)}{dt} \overset{!}{=} 0$. Differentiating with respect to the upper integral boundary gives:

$$\frac{1}{\pi} \frac{2\pi N}{T} \frac{\sin z}{z} \overset{!}{=} 0 \tag{1.65}$$

or $z = l\pi$ with $l = 1, 2, 3, \ldots$. The first extreme on $t_1 = T/(2N)$ is a maximum, the second extreme at $t_2 = T/N$ is a minimum (as can easily be seen). The extremes get closer and closer to each other for $N \to \infty$. How big is $S_N(t_1)$? Insertion in (1.64) gives us the value of the "overshoot":

$$S_N(t_1) \to \frac{1}{\pi} \int\limits_0^\pi \frac{\sin z}{z} dz = \frac{1}{2} + 0.0895. \tag{1.66}$$

Using the same method we get the value of the "undershoot":

$$S_N(t_2) \to \frac{1}{\pi} \int\limits_0^{2\pi} \frac{\sin z}{z} dz = \frac{1}{2} - 0.048. \tag{1.67}$$

I bet you've noticed that, in the approximation of N big and t small, the value of the overshoot or undershoot doesn't depend on N at all any more. Therefore it doesn't make sense to make N as big as possible, the overshoots and undershoots will settle at values of $+0.0895$ and -0.048 and stay there. We could still show that the extremes decrease monotonically until $t = T/4$; thereafter they'll be mirrored and increase (cf. Fig. 1.14). Now what about our mean squared error for $N \to \infty$? The answer is simple: the mean squared error approaches 0 for $N \to \infty$, though the overshoots and undershoots stay. That's the trick: as the extremes get closer and closer to each other, the area covered by the overshoots and the undershoots with the function $f(t) = 1/2$ $(t > 0)$ approaches 0 all the same. Integration will only come up with areas of measure 0 (I'm sure I've committed at least a venial sin by putting it this way). The moral of the story: a kink in the function (non-differentiability) lands us with an infinite Fourier series, and a step (discontinuity) gives us Gibbs' "ringing" to boot. In a nutshell: avoid steps wherever it's possible!

Playground

1.1 Very Speedy
A broadcasting station transmits on 100 MHz. Calculate the angular frequency ω and the period T for one complete oscillation. How far travels an electromagnetic pulse (or a light pulse!) in this time? Use the vacuum velocity of light $c \approx 3 \times 10^8$ m/s.

1.2 Totally Odd

Given is the function $f(t) = \cos(\pi t/2)$ for $0 < t \leq 1$ with periodic continuation. Plot this function. Is this function even, odd, or mixed? If it is mixed, decompose it into even and odd components and plot them.

1.3 Absolutely True

Calculate the complex Fourier coefficients C_k for $f(t) = \sin \pi t$ for $0 \leq t \leq 1$ with periodic continuation. Plot $f(t)$ with periodic continuation. Write down the first four terms in the series expansion.

1.4 Rather Complex

Calculate the complex Fourier coefficients C_k for $f(t) = 2 \sin(3\pi t/2) \cos(\pi t/2)$ for $0 \leq t \leq 1$ with periodic continuation. Plot $f(t)$.

1.5 Shiftily

Shift the function $f(t) = 2 \sin(3\pi t/2) \cos(\pi t/2) = \sin \pi t + \sin 2\pi t$ for $0 \leq t \leq 1$ with periodic continuation by $a = -1/2$ to the left and calculate the complex Fourier coefficient C_k. Plot the shifted $f(t)$ and its decomposition into first and second parts and discuss the result.

1.6 Cubed

Calculate the complex Fourier coefficients C_k for $f(t) = \cos^3 2\pi t$ for $0 \leq t \leq 1$ with periodic continuation. Plot this function. Now use (1.5) and the Second Shifting Rule to check your result.

1.7 Tackling Infinity

Derive the result for the infinite series $\sum_{k=1}^{\infty} 1/k^4$ using Parseval's theorem.
   Hint: Instead of the triangular function try a parabola!

1.8 Smoothly

Given is the function $f(t) = [1 - (2t)^2]^2$ for $-1/2 \leq t \leq 1/2$ with periodic continuation. Use (1.63) and argue how the Fourier coefficients C_k must depend on k. Check it by calculating the C_k directly.

Chapter 2
Continuous Fourier Transformation

Abstract This chapter deals with the mapping of arbitrary functions in the time domain to Fourier transformed functions in the frequency domain. The so-called δ-function is introduced and forward and inverse transformations are defined and illustrated with examples. The polar representation of the Fourier transform is given and shifting rules are discussed. Convolution, cross-correlation, and autocorrelation with Parseval's theorem are illustrated with examples. It concludes with a discussion of pitfalls and truncation errors.

In this chapter we relax the requirement of periodicity of the function $f(t)$. Hence, instead of discrete Fourier coefficients we end up with the continuous function $F(\omega)$. The integration interval is the entire real axis $(-\infty, +\infty)$.

Mapping of an *Arbitrary* Function $f(t)$ to the Fourier-Transformed Function $F(\omega)$

2.1 Continuous Fourier Transformation

We'll look at what happens at the transition from a series- to an integral-representation:

$$\text{Series:} \qquad C_k = \frac{1}{T} \int\limits_{-T/2}^{+T/2} f(t)\mathrm{e}^{-2\pi \mathrm{i} k t / T}\,\mathrm{d}t. \qquad (2.1)$$

$$\text{Now:} \quad T \to \infty \quad \omega_k = \frac{2\pi k}{T} \quad \to \quad \omega,$$

$$\qquad\qquad \text{discrete} \qquad\qquad \text{continuous}$$

$$\lim_{T \to \infty} (T C_k) \to \int\limits_{-\infty}^{+\infty} f(t)\mathrm{e}^{-\mathrm{i}\omega t}\,\mathrm{d}t. \qquad (2.2)$$

Before we get into the definition of the Fourier transformation, we have to do some homework.

© Springer International Publishing Switzerland 2015

T. Butz, *Fourier Transformation for Pedestrians*, Undergraduate Lecture Notes in Physics, DOI 10.1007/978-3-319-16985-9_2

2.1.1 Even and Odd Functions

A function is called even, if:
$$f(-t) = f(t). \tag{2.3}$$

A function is called odd, if:
$$f(-t) = -f(t). \tag{2.4}$$

Any general function may be split into an even and an odd part. We've heard that before, at the beginning of Chap. 1, and of course it's true whether the function $f(t)$ is periodic or not.

2.1.2 The δ-Function

The δ-function is a distribution,[1] not a function. In spite of that, it's always called δ-function. Its value is zero anywhere except when its argument is equal to 0. In this case it is ∞. If you think that's too steep or pointed for you, you may prefer a different definition:

$$\delta(t) = \lim_{a \to \infty} f_a(t)$$
$$\text{with } f_a(t) = \begin{cases} a \text{ for } -\dfrac{1}{2a} \le t \le \dfrac{1}{2a} \\ 0 \text{ else} \end{cases} . \tag{2.5}$$

Now we have a pulse for the duration of $-1/2a \le t \le 1/2a$ with height a and keep diminishing the width of the pulse while keeping the area unchanged (normalised to 1), viz. the height goes up while the width gets smaller. That's the reason why the δ-function often is also called impulse. At the end of the previous chapter we already had heard about a representation of the δ-function: Dirichlet's kernel for $N \to \infty$. If we restrict things to the basis interval $-\pi \le t \le +\pi$, we get:

$$\int_{-\pi}^{+\pi} D_N(x)\mathrm{d}x = \pi, \text{ independent of } N, \tag{2.6}$$

---

[1]Generalised function. The theory of distributions is an important basis of modern analysis, and impossible to understand without additional reading. A more in-depth treatment of its theory, however, is not required for the applications in this book.

and thus:

$$\frac{1}{\pi} \lim_{N \to \infty} \int_{-\pi}^{+\pi} f(t) D_N(t) \mathrm{d}t = f(0). \tag{2.7}$$

In the same way, the δ-function "picks" the integrand where the latter's argument is 0 during integration (we always have to integrate over the δ-function!):

$$\int_{-\infty}^{+\infty} f(t) \delta(t) \mathrm{d}t = f(0). \tag{2.8}$$

Another representation for the δ-function, which we'll frequently use, is:

$$\delta(\omega) = \frac{1}{2\pi} \int_{-\infty}^{+\infty} \mathrm{e}^{\mathrm{i}\omega t} \mathrm{d}t. \tag{2.9}$$

Purists may multiply the integrand with a damping-factor, for example $\mathrm{e}^{-\alpha|t|}$, and then introduce $\lim_{\alpha \to 0}$. This won't change the fact that everything gets "oscillated" or averaged away for all frequencies $\omega \neq 0$ (venial sin: let's think in whole periods for once!), whereas for $\omega = 0$ integration will be over the integrand 1 from $-\infty$ to $+\infty$, i.e. the result will have to be ∞.

2.1.3 Forward and Inverse Transformation

Let's define:

Definition 2.1 (*Forward transformation*)

$$F(\omega) = \int_{-\infty}^{+\infty} f(t) \mathrm{e}^{-\mathrm{i}\omega t} \mathrm{d}t. \tag{2.10}$$

Definition 2.2 (*Inverse transformation*)

$$f(t) = \frac{1}{2\pi} \int_{-\infty}^{+\infty} F(\omega) \mathrm{e}^{+\mathrm{i}\omega t} \mathrm{d}\omega. \tag{2.11}$$

Caution:

i. In the case of the forward transformation, there is a minus sign in the exponent
   (cf. (1.27)), in the case of the inverse transformation, this is a plus sign.
ii. In the case of the inverse transformation, $1/2\pi$ is in front of the integral, contrary
   to the forward transformation.

The asymmetric aspect of the formulas has tempted many scientists to introduce
other definitions, for example to write a factor $1/\sqrt{(2\pi)}$ for forward as well as inverse
transformation. That's no good, as the definition of the average $F(0) = \int_{-\infty}^{+\infty} f(t)dt$
would be affected. Weaver's representation is correct, though not widely used:

$$\text{Forward transformation:} \quad F(\nu) = \int\limits_{-\infty}^{+\infty} f(t)e^{-2\pi i\nu t}\,dt,$$

$$\text{Inverse transformation:} \quad f(t) = \int\limits_{-\infty}^{+\infty} F(\nu)e^{2\pi i\nu t}\,d\nu.$$

Weaver, as can be seen, doesn't use the angular frequency ω, but rather the frequency ν. This effectively made the formulas look symmetrical, though it saddles us
with many factors 2π in the exponent. We'll stick to the definitions (2.10) and (2.11).
   We now want to demonstrate, that the inverse transformation returns us to the
original function. For the forward transformation, we often will use FT($f(t)$), and
for the inverse transformation we'll use $\text{FT}^{-1}(F(\omega))$. We'll start with the inverse
transformation and insert:

$$f(t) = \frac{1}{2\pi} \int\limits_{-\infty}^{+\infty} F(\omega)e^{i\omega t}\,d\omega = \frac{1}{2\pi} \int\limits_{-\infty}^{+\infty} d\omega \int\limits_{-\infty}^{+\infty} f(t')e^{-i\omega t'}e^{i\omega t}\,dt'$$

$$= \frac{1}{2\pi} \int\limits_{-\infty}^{+\infty} f(t')dt' \int\limits_{-\infty}^{+\infty} e^{i\omega(t-t')}\,d\omega$$

$$\text{interchange integration} \tag{2.12}$$

$$= \int\limits_{-\infty}^{+\infty} f(t')\delta(t-t')dt' = f(t)\,.$$

Q.e.d.[2] Here we have used (2.8) and (2.9). For $f(t) = 1$ we get:

$$\text{FT}(\delta(t)) = 1. \tag{2.13}$$

---

[2]In Latin: "quod erat demonstrandum", "what we've set out to prove".

The impulse therefore requires all frequencies with unity amplitude for its Fourier representation ("white" spectrum). Conversely:

$$\text{FT}(1) = 2\pi\delta(\omega). \tag{2.14}$$

The constant 1 can be represented by a single spectral component, viz. $\omega = 0$. No others occur. As we have integrated from $-\infty$ to $+\infty$, naturally an $\omega = 0$ will also result in infinity for intensity.

We realise the dual character of the forward and inverse transformations: a very slowly varying function $f(t)$ will have a very high spectral density for very small frequencies; the spectral density will go down quickly and rapidly approaches 0. Conversely, a quickly varying function $f(t)$ will show spectral density over a very wide frequency range: Fig. 2.1 explains this once again.

Let's discuss a few examples now.

Example 2.1 ("Rectangle, even")

$$f(t) = \begin{cases} 1 \text{ for} & -T/2 \le t \le T/2 \\ 0 \text{ else} \end{cases}.$$

$$F(\omega) = 2 \int_{0}^{T/2} \cos\omega t\, dt = T\frac{\sin(\omega T/2)}{\omega T/2}. \tag{2.15}$$

The imaginary part is 0, as $f(t)$ is even. The Fourier transformation of a rectangular function therefore is of the type $\frac{\sin x}{x}$. Some authors use the expression $\text{sinc}(x)$

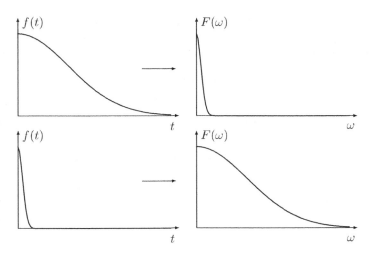

Fig. 2.1 A slowly-varying function has only low-frequency spectral components (*top*); a rapidly-falling function has spectral components spanning a wide range of frequencies (*bottom*)

Fig. 2.2 "Rectangular function" and Fourier transformation of type $\frac{\sin x}{x}$

for this case. What the "c" stands for, I don't know.[3] The "c" already has been "used up" when defining the complementary error-function $\text{erfc}(x) = 1 - \text{erf}(x)$. That's why we'd rather stick to $\frac{\sin x}{x}$. These functions $f(t)$ and $F(\omega)$ are shown in Fig. 2.2. They'll keep us busy for quite a while.

Keen readers would have spotted the following immediately: if we made the interval smaller and smaller, and did not fix $f(t)$ at 1 in return, but let it grow at the same rate as T decreases ("so the area under the curve stays constant"), then in $\lim_{T\to\infty}$ we would have a new representation of the δ-function. Again, we get the case where over- and undershoots on the one hand get closer to each other when T gets smaller, but on the other hand, their amplitude doesn't decrease. The shape $\frac{\sin x}{x}$ will stay the same. As we're already familiar with Gibbs' phenomenon in the case of steps, this naturally won't surprise us any more. Contrary to the discussion in Sect. 1.4.3 we don't have a periodic continuation of $f(t)$ beyond the integration interval, i.e. there are two steps (one up, one down). It's irrelevant that $f(t)$ on average isn't 0. It's important that for:

$$\omega \to 0 \qquad \sin(\omega T/2)/(\omega T/2) \to 1$$

(use l'Hospital's rule or $\sin x \approx x$ for small x).

Now, we calculate the Fourier transform of important functions. Let's start with the Gaussian.

Example 2.2 (The normalised Gaussian) The prefactor is chosen in such a way that the area is 1.

$$f(t) = \frac{1}{\sigma\sqrt{2\pi}}e^{-\frac{1}{2}\frac{t^2}{\sigma^2}}.$$

$$F(\omega) = \frac{1}{\sigma\sqrt{2\pi}}\int\limits_{-\infty}^{+\infty} e^{-\frac{1}{2}\frac{t^2}{\sigma^2}}e^{-i\omega t}\mathrm{d}t \qquad (2.16)$$

---

[3] It stands for "sinus cardinalis", but what is "cardinalis"? Has nothing to do with the catholic church, I guess.

$$= \frac{2}{\sigma\sqrt{2\pi}} \int\limits_{0}^{+\infty} e^{-\frac{1}{2}\frac{t^2}{\sigma^2}} \cos\omega t\, dt$$

$$= e^{-\frac{1}{2}\sigma^2\omega^2}.$$

Again, the imaginary part is 0, as $f(t)$ is even. The Fourier transform of a Gaussian results in another Gaussian. Note that the Fourier transform is not normalised to area 1. The $1/2$ occurring in the exponent is handy (could also have been absorbed into σ), as the following is true for this representation:

$$\sigma = \sqrt{2\ln 2} \times \text{HWHM (half width at half maximum} = \text{HWHM)} \qquad (2.17)$$
$$= 1.177 \times \text{HWHM}.$$

$f(t)$ has σ in the exponent's denominator, $F(\omega)$ in the numerator: the slimmer $f(t)$, the wider $F(\omega)$ and vice versa (cf. Fig. 2.3).

Example 2.3 (Bilateral exponential function)

$$f(t) = e^{-|t|/\tau}.$$

$$F(\omega) = \int\limits_{-\infty}^{+\infty} e^{-|t|/\tau} e^{-i\omega t}\, dt = 2\int\limits_{0}^{+\infty} e^{-t/\tau} \cos\omega t\, dt = \frac{2\tau}{1 + \omega^2\tau^2}. \qquad (2.18)$$

As $f(t)$ is even, the imaginary part is 0. The Fourier transform of the exponential function is a Lorentzian (cf. Fig. 2.4).

$$f(t) = \frac{1}{\sigma\sqrt{2\pi}} e^{-\frac{1}{2}\frac{t^2}{\sigma^2}} \qquad\qquad F(\omega) = e^{-\frac{1}{2}\sigma^2\omega^2}$$

Fig. 2.3 Gaussian and Fourier transform (=equally a Gaussian)

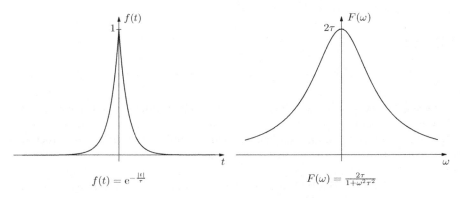

$$f(t) = \mathrm{e}^{-\frac{|t|}{\tau}} \qquad\qquad F(\omega) = \frac{2\tau}{1+\omega^2\tau^2}$$

Fig. 2.4 Bilateral exponential function and Fourier transformation (=Lorentzian)

Example 2.4 (Unilateral exponential function)

$$f(t) = \begin{cases} \mathrm{e}^{-\lambda t} & \text{for } t \geq 0 \\ 0 & \text{else} \end{cases}. \tag{2.19}$$

$$F(\omega) = \int\limits_0^\infty \mathrm{e}^{-\lambda t}\mathrm{e}^{-\mathrm{i}\omega t}\,\mathrm{d}t = \left.\frac{\mathrm{e}^{-(\lambda+\mathrm{i}\omega)t}}{-(\lambda+\mathrm{i}\omega)}\right|_0^{+\infty} \tag{2.20}$$

$$= \frac{1}{\lambda+\mathrm{i}\omega} = \frac{\lambda}{\lambda^2+\omega^2} + \frac{-\mathrm{i}\omega}{\lambda^2+\omega^2}. \tag{2.21}$$

(*Sorry*: When integrating in the complex plane, we really should have used the Residue Theorem[4] instead of integrating in a rather cavalier fashion. The result, however, is correct all the same.)

$F(\omega)$ is complex, as $f(t)$ is neither even nor odd. We now can write the real and the imaginary parts separately. The real part has a Lorentzian shape we're familiar with by now, and the imaginary part has a dispersion shape. Often the so-called polar representation is used, too, so we'll deal with that one in the next section.

Examples in physics: the damped oscillation that is used to describe the emission of a particle (for example a photon, a γ-quantum) from an excited nuclear state with a lifetime of τ (meaning, that the excited state depopulates according to $\mathrm{e}^{-t/\tau}$), results in a Lorentzian-shaped emission-line. Exponential relaxation processes will result in Lorentzian-shaped spectral-lines, for example in the case of nuclear magnetic resonance.

---

[4]The Residue Theorem is part of the theory of functions of complex variables.

2.1.4 Polar Representation of the Fourier Transform

Every complex number $z = a + ib$ can be represented in the complex plane by its magnitude and phase φ (Fig. 2.5):

$$z = a + ib = \sqrt{a^2 + b^2}\, e^{i\varphi} \text{ with } \tan\varphi = b/a.$$

This allows us to represent the Fourier transform of the "unilateral" exponential function as in Fig. 2.6.

Alternatively to the polar representation we can also represent the real and imaginary parts separately (cf. Fig. 2.7).

Please note that $|F(\omega)|$ is no Lorentzian! If you want to "stick" to this property, you better represent the square of the magnitude: $|F(\omega)|^2 = 1/(\lambda^2 + \omega^2)$ is a Lorentzian again. This representation is often also called the power representation: $|F(\omega)|^2 = (\text{real part})^2 + (\text{imaginary part})^2$. The phase goes to 0 at the maximum of $|F(\omega)|$, i.e. when "in resonance".

Warning: The representation of the magnitude as well as of the squared magnitude does away with the *linearity* of the Fourier transformation!

Finally, let's try out the inverse transformation and find out how we return to the "unilateral" exponential function (the Fourier transform didn't look all that "unilateral"!).

Fig. 2.5 Polar representation of a complex number $z = a + ib$

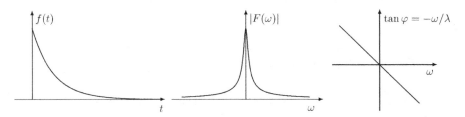

Fig. 2.6 Unilateral exponential function, magnitude of the Fourier transform and phase (imaginary part/real part)

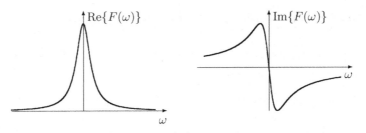

Fig. 2.7 Real part and imaginary part of the Fourier transform of a unilateral exponential function

$$f(t) = \frac{1}{2\pi} \int\limits_{-\infty}^{+\infty} \frac{\lambda - i\omega}{\lambda^2 + \omega^2} e^{i\omega t} d\omega$$

$$= \frac{1}{2\pi} \left\{ 2\lambda \int\limits_{0}^{+\infty} \frac{\cos \omega t}{\lambda^2 + \omega^2} d\omega + 2 \int\limits_{0}^{+\infty} \frac{\omega \sin \omega t}{\lambda^2 + \omega^2} d\omega \right\} \qquad (2.22)$$

$$= \frac{1}{\pi} \left\{ \frac{\pi}{2} e^{-|\lambda t|} \pm \frac{\pi}{2} e^{-|\lambda t|} \right\}, \text{ where } \begin{array}{l} \text{``+'' for } t \geq 0 \\ \text{``-'' for } t < 0 \end{array} \text{ is valid}$$

$$= \begin{cases} e^{-\lambda t} \text{ for } t \geq 0 \\ 0 \qquad \text{else} \end{cases}.$$

2.2 Theorems and Rules

2.2.1 Linearity Theorem

For completeness' sake, once again:

$$\begin{array}{c} f(t) \leftrightarrow F(\omega), \\ g(t) \leftrightarrow G(\omega), \\ a \cdot f(t) + b \cdot g(t) \leftrightarrow a \cdot F(\omega) + b \cdot G(\omega). \end{array} \qquad (2.23)$$

2.2.2 The First Shifting Rule

We already know: shifting in the time domain means modulation in the frequency domain:

$$\begin{array}{c} f(t) \leftrightarrow F(\omega), \\ f(t - a) \leftrightarrow F(\omega) e^{-i\omega a}. \end{array} \qquad (2.24)$$

The proof is quite simple.

Example 2.5 ("Rectangular function")

$$f(t) = \begin{cases} 1 \text{ for } T/2 \le t \le T/2 \\ 0 \text{ else} \end{cases}.$$

$$F(\omega) = T\frac{\sin(\omega T/2)}{\omega T/2}. \tag{2.25}$$

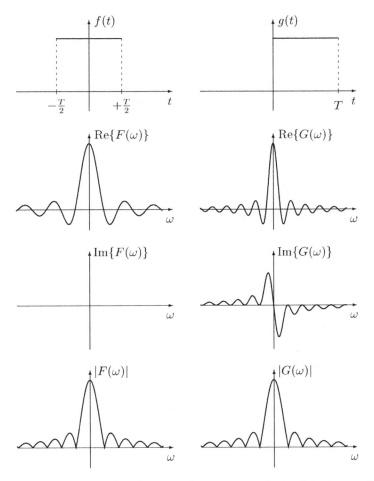

Fig. 2.8 "Rectangular function", real part, imaginary part, magnitude of Fourier transform (*left from top to bottom*); for the "rectangular function", shifted to the *right* by $T/2$ (*right from top to bottom*)

Now we shift the rectangle $f(t)$ by $a = T/2 \to g(t)$, and then get (see Fig. 2.8):

$$
\begin{aligned}
G(\omega) &= T\frac{\sin(\omega T/2)}{\omega T/2}e^{-i\omega T/2} \\
&= T\frac{\sin(\omega T/2)}{\omega T/2}(\cos(\omega T/2) - i\sin(\omega T/2)).
\end{aligned}
\tag{2.26}
$$

The real part gets modulated with $\cos(\omega T/2)$. The imaginary part which before was 0, now is unequal to 0 and "complements" the real part exactly, so $|F(\omega)|$ stays the same. Equation (2.24) contains "only" a phase factor $e^{-i\omega a}$, which is irrelevant as far as the magnitude is concerned. As long as you only look at the power spectrum, you may shift the function $f(t)$ along the time-axis as much as you want: you won't notice any effect. In the phase of the polar representation, however, you'll see the shift again:

$$
\tan\varphi = \frac{\text{imaginary part}}{\text{real part}} = -\frac{\sin(\omega T/2)}{\cos(\omega T/2)} = -\tan(\omega T/2)
\tag{2.27}
$$
$$
\text{or } \varphi = -\omega T/2.
$$

Don't worry about the phase φ overshooting $\pm\pi/2$.

2.2.3 The Second Shifting Rule

We already know: a modulation in the time domain results in a shift in the frequency domain:

$$
\begin{aligned}
f(t) &\leftrightarrow F(\omega), \\
f(t)e^{i\omega_0 t} &\leftrightarrow F(\omega - \omega_0).
\end{aligned}
\tag{2.28}
$$

If you prefer real modulations, you may write:

$$
\begin{aligned}
\text{FT}(f(t)\cos\omega_0 t) &= \frac{F(\omega + \omega_0) + F(\omega - \omega_0)}{2}, \\
\text{FT}(f(t)\sin\omega_0 t) &= i\frac{F(\omega + \omega_0) - F(\omega - \omega_0)}{2}.
\end{aligned}
\tag{2.29}
$$

This follows from Euler's identity (1.22) straight away.

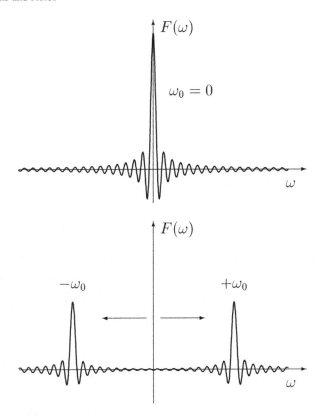

Fig. 2.9 Fourier transform of $g(t) = \cos \omega t$ in the interval $-T/2 \leq t \leq T/2$

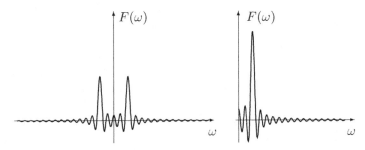

Fig. 2.10 Superposition of $\frac{\sin x}{x}$ sidelobes at small frequencies for negative and positive (*left*) and positive frequencies only (*right*)

Example 2.6 ("Rectangular function")

$$f(t) = \begin{cases} 1 \text{ for } -T/2 \le t \le +T/2 \\ 0 \text{ else} \end{cases}.$$

$$F(\omega) = T \frac{\sin(\omega T/2)}{\omega T/2} \qquad \text{(cf. (2.15))}$$

and

$$g(t) = \cos \omega_0 t. \qquad (2.30)$$

Using $h(t) = f(t) \cdot g(t)$ and the Second Shifting Rule we get:

$$H(\omega) = \frac{T}{2} \left\{ \frac{\sin[(\omega + \omega_0)T/2]}{(\omega + \omega_0)T/2} + \frac{\sin[(\omega - \omega_0)T/2]}{(\omega - \omega_0)T/2} \right\}. \qquad (2.31)$$

This means: the Fourier transform of the function $\cos \omega_0 t$ within the interval $-T/2 \le t \le T/2$ (and outside equal to 0) consists of two frequency peaks, one at $\omega = -\omega_0$ and another one at $\omega = +\omega_0$. The amplitude naturally gets split evenly ("between brothers"). If we had $\omega_0 = 0$, then we'd get the central peak $\omega = 0$ once again; increasing ω_0 splits this peak into two peaks, moving to the left and the right (cf. Fig. 2.9).

If you don't like negative frequencies, you may flip the negative half-plane, so you'll only get *one* peak at $\omega = \omega_0$ with twice (that's the original) intensity.

Caution: For small frequencies ω_0 the sidelobes of the function $\frac{\sin x}{x}$ tend to "rub shoulders", meaning that they interfere with each other. Even flipping the negative half-plane won't help that. Figure 2.10 explains the problem.

2.2.4 Scaling Theorem

Similar to (1.41) the following is true:

$$f(t) \leftrightarrow F(\omega),$$

$$f(at) \leftrightarrow \frac{1}{|a|} F\left(\frac{\omega}{a}\right). \qquad (2.32)$$

Proof (Scaling) Analogously to (1.41) with the difference that here we cannot stretch or compress the interval limits $\pm\infty$:

$$
\begin{aligned}
F(\omega)^{\text{new}} &= \frac{1}{T} \int\limits_{-\infty}^{+\infty} f(at)e^{-i\omega t}\, \mathrm{d}t \\
&= \frac{1}{T} \int\limits_{-\infty}^{+\infty} f(t')e^{-i\omega t'/a}\frac{1}{a}\mathrm{d}t' \quad \text{with } t' = at \\
&= \frac{1}{|a|} F(\omega)^{\text{old}} \quad\quad\quad \text{with } \omega = \frac{\omega^{\text{old}}}{a}. \quad \square
\end{aligned}
$$

Here, we tacitly assumed $a > 0$. For $a < 0$ we would get a minus sign in the prefactor; however, we would also have to interchange the integration limits and thus get together the factor $\frac{1}{|a|}$. This means: stretching (compressing) the time-axis results in the compression (stretching) of the frequency-axis. For the special case $a = -1$ we get:

$$
\begin{aligned}
f(t) &\rightarrow F(\omega), \\
f(-t) &\rightarrow F(-\omega).
\end{aligned}
\tag{2.33}
$$

Therefore turning around the time axis ("looking into the past") results in turning around the frequency axis. This profound secret will stay hidden to all those unable to think in anything but positive frequencies.

2.3 Convolution, Cross Correlation, Autocorrelation, Parseval's Theorem

2.3.1 Convolution

The convolution of a function $f(t)$ with another function $g(t)$ means:

Definition 2.3 (*Convolution*)

$$
f(t) \otimes g(t) \equiv \int\limits_{-\infty}^{+\infty} f(\xi)g(t - \xi)\mathrm{d}\xi.
\tag{2.34}
$$

Please note there is a minus sign in the argument of $g(t)$. The convolution is commutative, distributive, and associative. This means:

$$\text{commutative:} \quad f(t) \otimes g(t) = g(t) \otimes f(t).$$

Here we have to take into account the sign!

Proof (Convolution, commutative) Substituting the integration variables:

$$f(t) \otimes g(t) = \int\limits_{-\infty}^{+\infty} f(\xi)g(t-\xi)\mathrm{d}\xi = \int\limits_{-\infty}^{+\infty} g(\xi')f(t-\xi')\mathrm{d}\xi'$$

$$\text{with } \xi' = t - \xi. \quad \square$$

$$\text{Distributive:} \quad f(t) \otimes (g(t) + h(t)) = f(t) \otimes g(t) + f(t) \otimes h(t)$$

(Proof: *Linear operation!*).

$$\text{Associative:} \quad f(t) \otimes (g(t) \otimes h(t)) = (f(t) \otimes g(t)) \otimes h(t)$$

(the convolution sequence doesn't matter; proof: double integral with interchange of integration sequence).

Example 2.7 (Convolution of a "rectangular function" with another "rectangular function") We want to convolve the "rectangular function" $f(t)$ with another "rectangular function" $g(t)$:

$$f(t) = \begin{cases} 1 \text{ for } -T/2 \le t \le T/2 \\ 0 \text{ else} \end{cases},$$

$$g(t) = \begin{cases} 1 \text{ for } 0 \le t \le T \\ 0 \text{ else} \end{cases}.$$

$$h(t) = f(t) \otimes g(t). \tag{2.35}$$

According to the definition in (2.34) we have to mirror $g(t)$ (minus sign in front of ξ). Then we shift $g(t)$ and calculate the overlap (cf. Fig. 2.11).

We get the first overlap for $t = -T/2$ and the last one for $t = +3T/2$ (cf. Fig. 2.12).

At the limits, where $t = -T/2$ and $t = +3T/2$, we start and finish with an overlap of 0, the maximum overlap occurs at $t = +T/2$: there the two rectangles are exactly on top of each other (or below each other?). The integral then is exactly T; in between the integral rises/falls at a linear rate (cf. Fig. 2.13).

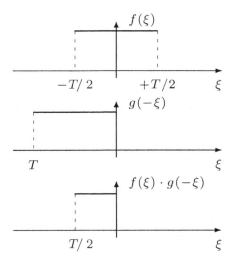

Fig. 2.11 "Rectangular function" $f(\xi)$, mirrored rectangular function $g(-\xi)$, overlap (*from top to bottom*). The area of the overlap gives the convolution integral

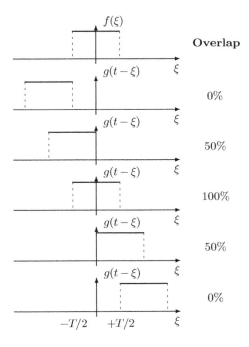

Fig. 2.12 Illustration of the convolution process of $f(t)$ and $g(t)$ with $t = -T/2, 0, +T/2, +T, +3T/2$ (*from top to bottom*)

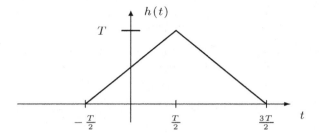

Fig. 2.13 Convolution $h(t) = f(t) \otimes g(t)$

Please note the following: the interval, where $f(t) \otimes g(t)$ is unequal to 0, now is twice as big: $2T$! If we had defined $g(t)$ symmetrically around 0 in the first place (I didn't want to do that, so we can't forget the mirroring!), then also $f(t) \otimes g(t)$ would be symmetrical around 0. In this case we would have convolved $f(t)$ with itself.

Now to a more useful example: let's take a pulse that looks like a "unilateral" exponential function:

$$f(t) = \begin{cases} e^{-t/\tau} & \text{for } t \geq 0 \\ 0 & \text{else} \end{cases}. \tag{2.36}$$

Any device that delivers pulses as a function of time, has a finite rise-/decay-time, which for simplicity's sake we'll assume to be a Gaussian (see Fig. 2.14):

$$g(t) = \frac{1}{\sigma\sqrt{2\pi}} e^{-\frac{1}{2}\frac{t^2}{\sigma^2}}. \tag{2.37}$$

That is how our device would represent a δ-function—we can't get sharper than that. The function $g(t)$ therefore is the device's resolution function, which we'll have to use for the convolution of *all* signals we want to record. An example would be the bandwidth of an oscilloscope. We then need:

$$S(t) = f(t) \otimes g(t), \tag{2.38}$$

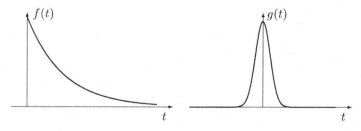

Fig. 2.14 Illustration of convolution: the Gaussian will be shifted over the unilateral exponential-function

where $S(t)$ is the experimental, "smeared" signal. It's obvious that the rise at $t = 0$ will not be as steep, and the peak of the exponential function will get "ironed out". We'll have to take a closer look:

$$S(t) = \frac{1}{\sigma\sqrt{2\pi}} \int\limits_0^{+\infty} e^{-\xi/\tau} e^{-\frac{1}{2}\frac{(t-\xi)^2}{\sigma^2}} d\xi$$

$$= \frac{1}{\sigma\sqrt{2\pi}} e^{-\frac{1}{2}\frac{t^2}{\sigma^2}} \int\limits_0^{+\infty} \exp \underbrace{\left[-\frac{\xi}{\tau} + \frac{t\xi}{\sigma^2} - \frac{1}{2}\xi^2/\sigma^2 \right]}_{\text{form quadratic complement}} d\xi$$

$$= \frac{1}{\sigma\sqrt{2\pi}} e^{-\frac{1}{2}\frac{t^2}{\sigma^2}} e^{\frac{t^2}{2\sigma^2}} e^{-\frac{t}{\tau}} e^{\frac{\sigma^2}{2\tau^2}} \int\limits_0^{+\infty} e^{-\frac{1}{2\sigma^2}\left(\xi - \left(t - \frac{\sigma^2}{\tau}\right)\right)^2} d\xi \qquad (2.39)$$

$$= \frac{1}{\sigma\sqrt{2\pi}} e^{-\frac{t}{\tau}} e^{+\frac{\sigma^2}{2\tau^2}} \int\limits_{-(t-\sigma^2/\tau)}^{+\infty} e^{-\frac{1}{2\sigma^2}\xi'^2} d\xi' \quad \text{with } \xi' = \xi - \left(t - \frac{\sigma^2}{\tau}\right)$$

$$= \frac{1}{2} e^{-\frac{t}{\tau}} e^{+\frac{\sigma^2}{2\tau^2}} \operatorname{erfc}\left(\frac{\sigma}{\sqrt{2}\tau} - \frac{t}{\sigma\sqrt{2}} \right).$$

Here, $\operatorname{erfc}(x) = 1 - \operatorname{erf}(x)$ is the complementary error-function with the defining equation:

$$\operatorname{erf}(x) = \frac{2}{\sqrt{\pi}} \int\limits_0^x e^{-t^2} dt. \qquad (2.40)$$

The functions $\operatorname{erf}(x)$ and $\operatorname{erfc}(x)$ are shown in Fig. 2.15.

The function $\operatorname{erfc}(x)$ represents a "smeared" step. Together with the factor $1/2$ the height of the step is just 1. As the time in the argument of $\operatorname{erfc}(x)$ in (2.39) has a negative sign, the step of Fig. 2.15 is mirrored and also shifted by $\sigma/\sqrt{2}\tau$. Figure 2.16 shows the result of the convolution of the exponential function with the Gaussian.

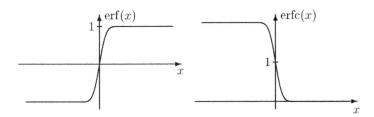

Fig. 2.15 The functions $\operatorname{erf}(x)$ and $\operatorname{erfc}(x)$

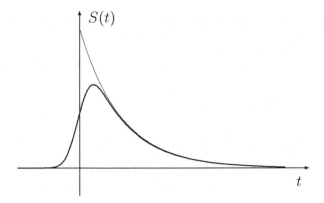

Fig. 2.16 Result of the convolution of a unilateral exponential function with a Gaussian. Exponential function without convolution (*thin line*)

The following properties immediately stand out:

i. the finite time resolution ensures that there also is a signal at negative times, whereas it was 0 before convolution,

ii. the maximum is not at $t = 0$ any more,

iii. what can't be seen straight away, yet is easy to grasp, is the following: the center of gravity of the exponential function, which was at $t = \tau$, doesn't get shifted at all upon convolution. An *even* function won't shift the center of gravity! Have a go and check it out!

It's easy to remember the shape of the curve in Fig. 2.16. Start out with the exponential function with a "90°-vertical cliff", and then dump "gravel" to the left and to the right of it (equal quantities! it's an even function!): that's how you get the gravel-heap for $t < 0$, demolish the peak and make sure there's also a gravel-heap for $t > 0$, that slowly gets thinner and thinner. Indeed, the influence of the step will become less and less important if times get larger and larger, i.e.:

$$\frac{1}{2}\mathrm{erfc}\left(\frac{\sigma}{\sqrt{2\tau}} - \frac{t}{\sigma\sqrt{2}}\right) \to 1 \qquad \text{for } t \gg \frac{\sigma^2}{\tau}, \tag{2.41}$$

and only the unchanged $\mathrm{e}^{-t/\tau}$ will remain, however with the constant factor $\mathrm{e}^{+\frac{\sigma^2}{2\tau^2}}$. This factor is always >1 because we always have more "gravel" poured downwards than upwards.

Now we prove the extremely important Convolution Theorem:

$$\begin{aligned}
f(t) &\leftrightarrow F(\omega), \\
g(t) &\leftrightarrow G(\omega), \\
h(t) = f(t) \otimes g(t) &\leftrightarrow H(\omega) = F(\omega) \cdot G(\omega),
\end{aligned} \tag{2.42}$$

i.e., the *convolution integral* becomes, through Fourier transformation, a *product* of the Fourier-transformed ones.

Proof (Convolution Theorem)

$$H(\omega) = \iint f(\xi)g(t-\xi)d\xi \times e^{-i\omega t}dt$$

$$= \int f(\xi)e^{-i\omega\xi}\left[\int g(t-\xi)e^{-i\omega(t-\xi)}dt\right]d\xi$$

$$\uparrow \qquad \text{expanded} \qquad \uparrow \qquad\qquad (2.43)$$

$$= \int f(\xi)e^{-i\omega\xi}d\xi \times G(\omega)$$

$$= F(\omega) \times G(\omega). \quad \square$$

In the step before the last one, we substituted $t' = t-\xi$. The integration boundaries $\pm\infty$ did not change by doing that, and $G(\omega)$ does not depend on ξ.

The inverse Convolution Theorem then is:

$$f(t) \leftrightarrow F(\omega),$$
$$g(t) \leftrightarrow G(\omega), \qquad\qquad (2.44)$$
$$h(t) = f(t) \cdot g(t) \leftrightarrow H(\omega) = \tfrac{1}{2\pi}F(\omega) \otimes G(\omega).$$

Proof (Inverse Convolution Theorem)

$$H(\omega) = \int f(t)g(t)e^{-i\omega t}dt$$

$$= \int \left(\frac{1}{2\pi}\int F(\omega')e^{+i\omega't}d\omega' \times \frac{1}{2\pi}\int G(\omega'')e^{+i\omega''t}d\omega''\right)e^{-i\omega t}dt$$

$$= \frac{1}{(2\pi)^2}\int F(\omega')\int G(\omega'')\underbrace{\int e^{i(\omega'+\omega''-\omega)t}dt}_{=2\pi\delta(\omega'+\omega''-\omega)} d\omega'd\omega''$$

$$= \frac{1}{2\pi}\int F(\omega')G(\omega-\omega')d\omega'$$

$$= \frac{1}{2\pi}F(\omega) \otimes G(\omega). \quad \square$$

Caution: Contrary to the Convolution Theorem (2.42), in (2.44) there is a factor of $1/2\pi$ in front of the convolution of the Fourier transforms.

A widely popular exercise is the de-convolution of data: the instruments' resolution function "smears out" the quickly varying functions, but we naturally want to reconstruct the data to what they would look like if the resolution function was infinitely good—provided we precisely knew the resolution function. In principle, that's a good idea—and thanks to the Convolution Theorem, not a problem: you

Fourier-transform the data, divide by the Fourier-transformed resolution function and transform it back. For practical applications it doesn't quite work that way. As in real life, we can't transform from $-\infty$ to $+\infty$, we need low-pass filters, in order not to get "swamped" with oscillations resulting from cut-off errors. Therefore the advantages of de-convolution are just as quickly lost as gained. Actually, the following is obvious: whatever got "smeared" by finite resolution, can't be reconstructed unambiguously. Imagine that a very pointed peak got eroded over millions of years, so there's only gravel left at its bottom. Try reconstructing the original peak from the debris around it! The result might be impressive from an artist's point of view, an artefact, but it hasn't got much to do with the original reality (unfortunately the word artefact has negative connotations among scientists).

Two useful examples for the Convolution Theorem:

Example 2.8 (Gaussian frequency distribution) Let's assume we have $f(t) = \cos\omega_0 t$, and the frequency ω_0 is not precisely defined, but is Gaussian distributed:

$$P(\omega) = \frac{1}{\sigma\sqrt{2\pi}}e^{-\frac{1}{2}\frac{\omega^2}{\sigma^2}}.$$

What we're measuring then is:

$$\tilde{f}(t) = \int\limits_{-\infty}^{+\infty} \frac{1}{\sigma\sqrt{2\pi}}e^{-\frac{1}{2}\frac{\omega^2}{\sigma^2}}\cos(\omega - \omega_0)t\mathrm{d}\omega, \qquad (2.45)$$

i.e. a convolution integral in ω_0. Instead of calculating this integral directly, we use the inverse of the Convolution Theorem (2.44), thus saving work and gaining higher enlightenment. But watch it! We have to handle the variables carefully. The time t in (2.45) has nothing to do with the Fourier transformation we need in (2.44). And the same is true for the integration variable ω. Therefore we rather use t_0 and ω_0 for the variable pairs in (2.44). We identify:

$$F(\omega_0) = \frac{1}{\sigma\sqrt{2\pi}}e^{-\frac{1}{2}\frac{\omega_0^2}{\sigma^2}}$$
$$\frac{1}{2\pi}G(\omega_0) = \cos\omega_0 t \qquad \text{or} \quad G(\omega_0) = 2\pi\cos\omega_0 t.$$

The inverse transformation of these functions using (2.11) gives us:

$$f(t_0) = \frac{1}{2\pi}e^{-\frac{1}{2}\sigma^2 t_0^2}$$

(cf. (2.16) for the inverse problem; don't forget the factor $1/2\pi$ when doing the inverse transformation!),

$$g(t_0) = 2\pi \left[\frac{\delta(t_0 - t)}{2} + \frac{\delta(t_0 + t)}{2} \right]$$

(cf. (2.9) for the inverse problem; use the First Shifting Rule (2.24); don't forget the factor $1/2\pi$ when doing the inverse transformation!).

Finally we get:

$$h(t_0) = e^{-\frac{1}{2}\sigma^2 t_0^2} \left[\frac{\delta(t_0 - t)}{2} + \frac{\delta(t_0 + t)}{2} \right].$$

Now the only thing left is to Fourier-transform $h(t_0)$. The integration over the δ-function actually is fun:

$$\widetilde{f}(t) \equiv H(\omega_0) = \int\limits_{-\infty}^{+\infty} e^{-\frac{1}{2}\sigma^2 t_0^2} \left[\frac{\delta(t_0 - t)}{2} + \frac{\delta(t_0 + t)}{2} \right] e^{-i\omega_0 t_0} dt_0$$

$$= e^{-\frac{1}{2}\sigma^2 t^2} \cos \omega_0 t.$$

Now, this was more work than we'd originally thought it would be. But look at what we've gained in insight!

This means: the convolution of a Gaussian distribution in the frequency domain results in exponential "damping" of the cosine term, where the damping happens to be the Fourier transform of the frequency distribution. This, of course, is due to the fact that we have chosen to use a cosine function (i.e. a basis function) for $f(t)$. $P(\omega)$ makes sure that oscillations for $\omega \neq \omega_0$ are slightly shifted with respect to each other, and will more and more superimpose each other destructively in the long run, averaging out to 0.

Example 2.9 (Lorentzian frequency distribution) Now naturally we'll know immediately what a convolution with a Lorentzian distribution:

$$P(\omega) = \frac{\sigma}{\pi} \frac{1}{\omega^2 + \sigma^2} \tag{2.46}$$

would do:

$$\widetilde{f}(t) = \int\limits_{-\infty}^{+\infty} \frac{\sigma}{\pi} \frac{1}{\omega^2 + \sigma^2} \cos(\omega - \omega_0) t d\omega,$$

$$h(t_0) = FT^{-1}(\widetilde{f}(t)) = e^{-\sigma t_0} \left[\frac{\delta(t_0 - t)}{2} + \frac{\delta(t_0 + t)}{2} \right]; \tag{2.47}$$

$$\widetilde{f}(t) = e^{-\sigma t} \cos \omega_0 t.$$

This is a damped wave. That's how we would describe the electric field of a Lorentz-shaped spectral line, sent out by an "emitter" with a life time of $1/\sigma$.

These examples are of fundamental importance to physics. Whenever we probe with plane waves, i.e. e^{iqx}, the answer we get is the Fourier transform of the respective distribution function of the object. A classical example is the elastic scattering of electrons at nuclei. Here, the form factor $F(q)$ is the Fourier transform of the distribution function of the nuclear charge density $\rho(x)$. The wave vector q is, apart from a prefactor, identical with the momentum.

Example 2.10 (Gaussian convolved with Gaussian) We perform a convolution of a Gaussian with σ_1 with another Gaussian with σ_2. As the Fourier transforms are Gaussians again—yet with σ_1^2 and σ_2^2 in the *numerator* of the exponent—it's immediately obvious that $\sigma_{total}^2 = \sigma_1^2 + \sigma_2^2$. Therefore, we get another Gaussian with geometric addition of the widths σ_1 and σ_2.

2.3.2 Cross Correlation

Sometimes we want to know if a measured function $f(t)$ has anything in common with another measured function $g(t)$. Cross correlation is ideally suited to that.

Definition 2.4 (*Cross correlation*)

$$h(t) = \int\limits_{-\infty}^{+\infty} f(\xi)g^*(t + \xi)\mathrm{d}\xi \equiv f(t) \star g(t). \qquad (2.48)$$

Watch it: Here, there is a plus sign in the argument of g, therefore we don't mirror $g(t)$. For even functions $g(t)$, this, however, doesn't matter.

The asterisk * means: complex conjugated. We may disregard it for real functions. The symbol $\star$ means: cross correlation, and is not to be confounded with $\otimes$ for convolution. Cross correlation is associative and distributive, yet *not* commutative. That's not only because of the complex-conjugated symbol, but mainly because of the plus sign in the argument of $g(t)$. Of course we want to convert the integral in the cross correlation to a product by using Fourier transformation.

$$f(t) \leftrightarrow F(\omega),$$
$$g(t) \leftrightarrow G(\omega), \qquad (2.49)$$
$$h(t) = f(t) \star g(t) \leftrightarrow H(\omega) = F(\omega)G^*(\omega).$$

Proof (Fourier transform of cross correlation)

$$H(\omega) = \iint f(\xi) g^*(t + \xi) d\xi \cdot e^{-i\omega t} dt$$

$$= \int f(\xi) \left[\int g^*(t + \xi) e^{-i\omega t} dt \right] d\xi$$

First Shifting Rule complex conjugated with $\xi = -a$ (2.50)

$$= \int f(\xi) G^*(+\omega) e^{-i\omega \xi} d\xi$$

$$= F(\omega) G^*(\omega). \quad \square$$

Here we used the following identity:

$$G(\omega) = \int g(t) e^{-i\omega t} dt$$

(take both sides complex conjugated)

$$G^*(\omega) = \int g^*(t) e^{i\omega t} dt \qquad (2.51)$$

$$G^*(-\omega) = \int g^*(t) e^{-i\omega t} dt$$

(ω to be replaced by $-\omega$).

The interpretation of (2.49) is simple: if the spectral densities of $f(t)$ and $g(t)$ are a good match, i.e. have much in common, then $H(\omega)$ will become large on average, and the cross correlation $h(t)$ will also be large, on average. Otherwise if $F(\omega)$ would be small e.g., where $G^*(\omega)$ is large and vice versa, so that there is never much left for the product $H(\omega)$. Then also $h(t)$ would be small, i.e. there is not much in common between $f(t)$ and $g(t)$.

A maybe somewhat extreme example is the technique of "Lock-in amplification", used to "dig up" small signals buried deeply in the noise. In this case we modulate the measured signal with a so-called carrier frequency, detect an extremely narrow spectral range—provided the desired signal does have spectral components in exactly this spectral width—and often additionally make use of phase information, too. Anything that doesn't correlate with the carrier frequency, gets discarded, so we're only left with the noise close to the working frequency.

2.3.3 Autocorrelation

The autocorrelation function is the cross correlation of a function $f(t)$ with itself. You may ask, for what purpose we'd want to check for what $f(t)$ has in common with $f(t)$. Autocorrelation, however, seems to attract many people in a magic manner. We often hear the view, that a signal full of noise can be turned into something really good by using the autocorrelation function, i.e. the signal-to-noise ratio would improve a lot. Don't you believe a word of it! We'll see why shortly.

Definition 2.5 (*Autocorrelation*)

$$h(t) = \int f(\xi) f^*(\xi + t) \mathrm{d}\xi. \tag{2.52}$$

We get:

$$\begin{aligned} f(t) &\leftrightarrow F(\omega), \\ h(t) = f(t) \star f(t) &\leftrightarrow H(\omega) = F(\omega) F^*(\omega) = |F(\omega)|^2. \end{aligned} \tag{2.53}$$

We may either use the Fourier transform $F(\omega)$ of a noisy function $f(t)$ and get angry about the noise in $F(\omega)$. Or we first form the autocorrelation function $h(t)$ from the function $f(t)$ and are then happy about the Fourier transform $H(\omega)$ of function $h(t)$. Normally, $H(\omega)$ does look a lot less noisy, indeed. Instead of doing it the roundabout way by using the autocorrelation function, we could have used the square of the magnitude of $F(\omega)$ in the first place. We all know, that a squared representation in the ordinate always pleases the eye, if we want to do cosmetics to a noisy spectrum. Big spectral components will grow when squared, small ones will get even smaller (cf. New Testament, Matthew 13:12: "For to him who has will more be given but from him who has not, even the little he has will be taken away."). But isn't it rather obvious that squaring doesn't change anything to the signal-to-noise ratio? In order to make it "look good", we pay the price of losing linearity.

Then, what is autocorrelation good for? A classical example comes from femtosecond measuring devices. A femtosecond is one part in a thousand trillion (US)—or a thousand billion (British)—of a second, not a particularly long time, indeed. Today, it is possible to produce such short laser pulses. How can we measure such short times? Using electronic stop-watches we can reach the range of 100 ps; hence, these "watches" are too slow by 5 orders of magnitude. Precision engineering does the job! Light travels in a femtosecond a distance of about 300 nm, i.e. about 1/100 of a hair diameter. Today you can buy positioning devices with nanometer precision. The trick: Split the laser pulse into two pulses, let them travel a slightly different optical length using mirrors, and combine them afterwards. The detector is an "optical coincidence" which yields an output only if both pulses overlap. By tuning the optical path (using the nanometer screw!) you can "shift" one pulse over the other, i.e. you perform a cross correlation of the pulse with itself (for purists: with its exact copy). The entire system is called autocorrelator.

2.3.4 Parseval's Theorem

The autocorrelation function also comes in handy for something else, namely for deriving Parseval's theorem. We start out with (2.52), insert especially $t = 0$, and get Parseval's theorem:

$$h(0) = \int |f(\xi)|^2 d\xi = \frac{1}{2\pi} \int |F(\omega)|^2 d\omega. \tag{2.54}$$

We get the second equal sign by inverse transformation of $|F(\omega)|^2$, in which for $t = 0$ the factor $e^{i\omega t}$ becomes unity.

Equation (2.54) states, that the "information content" of the function $f(x)$—defined as integral over the square of the magnitude—is just as large as the "information content" of its Fourier transform $F(\omega)$—defined as integral over the square of the magnitude of $F(\omega)$ divided by 2π. Let's check this out straight away using an example, namely our much-used "rectangular function"!

Example 2.11 ("Rectangular function")

$$f(t) = \begin{cases} 1 \text{ for } -T/2 \le t \le T/2 \\ 0 \text{ else} \end{cases}.$$

We get on the one hand:

$$\int\limits_{-\infty}^{+\infty} |f(t)|^2 dt = \int\limits_{-T/2}^{+T/2} dt = T$$

and on the other hand:

$$F(\omega) = T \frac{\sin(\omega T/2)}{\omega T/2}, \text{ thus}$$

$$\frac{1}{2\pi} \int\limits_{-\infty}^{+\infty} |F(\omega)|^2 d\omega = 2 \frac{T^2}{2\pi} \int\limits_0^{+\infty} \left[\frac{\sin(\omega T/2)}{\omega T/2} \right]^2 d\omega \tag{2.55}$$

$$= 2 \frac{T^2}{2\pi} \frac{2}{T} \int\limits_0^{+\infty} \left(\frac{\sin x}{x} \right)^2 dx = T$$

with $x = \omega T/2$.

It's easily understood that Parseval's theorem contains the squared magnitudes of both $f(t)$ and $F(\omega)$: anything unequal to 0 has information, regardless if it's positive or negative. The power spectrum is important, the phase doesn't matter. Of course we can use Parseval's theorem to calculate integrals. Let's simply take the last example for integration over $\left(\frac{\sin x}{x} \right)^2$. We need an integration table for that one, whereas integrating over 1, that's determining the area of a square, is elementary.

2.4 Fourier Transformation of Derivatives

When solving differential equations, we can make life easier using Fourier transformation. The derivative simply becomes a product:

$$f(t) \leftrightarrow F(\omega),$$
$$f'(t) \leftrightarrow i\omega F(\omega). \tag{2.56}$$

Proof (Fourier transformation of derivatives with respect to t) The abbreviation FT denotes the Fourier transformation:

$$\mathrm{FT}(f'(t)) = \int\limits_{-\infty}^{+\infty} f'(t)\mathrm{e}^{-i\omega t}\mathrm{d}t = f(t)\mathrm{e}^{-i\omega t}\Big|_{-\infty}^{+\infty} - (-i\omega) \int\limits_{-\infty}^{+\infty} f(t)\mathrm{e}^{-i\omega t}\mathrm{d}t$$

partial integration

$$= i\omega F(\omega). \quad \square$$

The first term in the partial integration is discarded, as $f(t) \to 0$ for $t \to \pm\infty$. Otherwise we would run into trouble with the integration, like we did at the end of Sect. 2.1.2. This game can go on:

$$\mathrm{FT}\left(\frac{\mathrm{d}f^n(t)}{\mathrm{d}^n t}\right) = (i\omega)^n F(\omega). \tag{2.57}$$

For negative n we may also use the formula for integration. We can also formulate in a simple way the derivative of a Fourier transform $F(\omega)$ with respect to the frequency ω:

$$\frac{\mathrm{d}F(\omega)}{\mathrm{d}\omega} = -i\mathrm{FT}(t f(t)). \tag{2.58}$$

Proof (Fourier transformation of derivatives with respect to ω)

$$\frac{\mathrm{d}F(\omega)}{\mathrm{d}\omega} = \int\limits_{-\infty}^{+\infty} f(t)\frac{\mathrm{d}}{\mathrm{d}\omega}\mathrm{e}^{-i\omega t}\mathrm{d}t = -i \int\limits_{-\infty}^{+\infty} f(t)t\mathrm{e}^{-i\omega t}\mathrm{d}t = -i\mathrm{FT}(t f(t)). \quad \square$$

Weaver [2] gives a neat example for the application of Fourier transformation:

Example 2.12 (Wave equation) The wave equation:

$$\frac{\mathrm{d}^2 u(x,t)}{\mathrm{d}t^2} = c^2 \frac{\mathrm{d}^2 u(x,t)}{\mathrm{d}x^2} \tag{2.59}$$

can be made into an oscillation equation using Fourier transformation of the local variable, which is much easier to solve. We assume:

$$U(\xi, t) = \int\limits_{-\infty}^{+\infty} u(x, t)e^{-i\xi x}dx.$$

Then we get:

$$\mathrm{FT}\left(\frac{d^2 u(x, t)}{dx^2}\right) = (i\xi)^2 U(\xi, t),$$

$$\mathrm{FT}\left(\frac{d^2 u(x, t)}{dt^2}\right) = \frac{d^2}{dt^2}U(\xi, t),$$

$$(2.60)$$

and all together:

$$\frac{d^2 U(\xi, t)}{dt^2} = -c^2\xi^2 U(\xi, t).$$

The solution of this equations is:

$$U(\xi, t) = P(\xi)\cos(c\xi t),$$

where $P(\xi)$ is the Fourier transform of the starting profile $p(x)$:

$$P(\xi) = \mathrm{FT}(p(x)) = U(\xi, 0).$$

The inverse transformation gives us two profiles propagating to the left and to the right:

$$u(x, t) = \frac{1}{2\pi} \int\limits_{-\infty}^{+\infty} P(\xi)\cos(c\xi t)e^{i\xi x}d\xi$$

$$= \frac{1}{2\pi}\frac{1}{2} \int\limits_{-\infty}^{+\infty} P(\xi)\left[e^{i\xi(x+ct)} + e^{i\xi(x-ct)}\right]d\xi \qquad (2.61)$$

$$= \frac{1}{2}p(x + ct) + \frac{1}{2}p(x - ct).$$

As we had no dispersion term in the wave equation, the profiles are conserved (cf. Fig. 2.17).

Fig. 2.17 Two starting profiles $p(x)$ propagating to the *left* and the *right* as solutions of the wave equation

2.5 Pitfalls

2.5.1 *"Turn 1 into 3"*

Just for fun, we'll get into magic now: let's take a unilateral exponential function:

$$f(t) = \begin{cases} e^{-\lambda t} \text{ for } t \geq 0 \\ 0 \quad \text{else} \end{cases}$$

$$\text{with } F(\omega) = \frac{1}{\lambda + i\omega} \tag{2.62}$$

$$\text{and } |F(\omega)|^2 = \frac{1}{\lambda^2 + \omega^2}.$$

We put this function (temporarily) on a unilateral "pedestal":

$$g(t) = \begin{cases} 1 \text{ for } t \geq 0 \\ 0 \text{ else} \end{cases}$$

$$\text{with } G(\omega) = \frac{1}{i\omega}. \tag{2.63}$$

We arrive at the Fourier transform of Heaviside's step function $g(t)$ from the Fourier transform for the exponential function for $\lambda \to 0$. We therefore have: $h(t) = f(t) + g(t)$. Because of the linearity of the Fourier transformation:

$$H(\omega) = \frac{1}{\lambda + i\omega} + \frac{1}{i\omega} = \frac{\lambda}{\lambda^2 + \omega^2} - \frac{i\omega}{\lambda^2 + \omega^2} - \frac{i}{\omega}. \tag{2.64}$$

This results in:

$$|H(\omega)|^2 = \left(\frac{\lambda}{\lambda^2 + \omega^2} - \frac{i\omega}{\lambda^2 + \omega^2} - \frac{i}{\omega}\right) \times \left(\frac{\lambda}{\lambda^2 + \omega^2} + \frac{i\omega}{\lambda^2 + \omega^2} + \frac{i}{\omega}\right)$$

$$= \frac{\lambda^2}{(\lambda^2 + \omega^2)^2} + \frac{1}{\omega^2} + \frac{\omega^2}{(\lambda^2 + \omega^2)^2} + \frac{2\omega}{(\lambda^2 + \omega^2)\omega}$$

$$= \frac{1}{\lambda^2 + \omega^2} + \frac{1}{\omega^2} + \frac{2}{\lambda^2 + \omega^2}$$

$$= \frac{3}{\lambda^2 + \omega^2} + \frac{1}{\omega^2}.$$

Now we return $|G(\omega)|^2 = 1/\omega^2$, i.e. the square of the Fourier transform of the pedestal, and have gained, compared to $|F(\omega)|^2$, a factor of 3. And we only had to temporarily "borrow" the pedestal to achieve that?! Of course (2.64) is correct. Returning $|G(\omega)|^2$ wasn't. We borrowed the interference term we got when squaring the magnitude, as well, and have to return it, too. This inference term amounts to just $2/(\lambda^2 + \omega^2)$.

Now let's approach the problem somewhat more academically. Assuming we have $h(t) = f(t) + g(t)$ with the Fourier transforms $F(\omega)$ and $G(\omega)$. We now use the polar representation:

$$F(\omega) = |F(\omega)|e^{i\varphi_f}$$
$$\text{and} \tag{2.65}$$
$$G(\omega) = |G(\omega)|e^{i\varphi_g}.$$

This gives us:

$$H(\omega) = |F(\omega)|e^{i\varphi_f} + |G(\omega)|e^{i\varphi_g}, \tag{2.66}$$

which is, due to the linearity of the Fourier transformation, entirely correct. However, if we want to calculate $|H(\omega)|^2$ (or the square root of it), we get:

$$|H(\omega)|^2 = \left(|F(\omega)|e^{i\varphi_f} + |G(\omega)|e^{i\varphi_g}\right)\left(|F(\omega)|e^{-i\varphi_f} + |G(\omega)|e^{-i\varphi_g}\right)$$
$$= |F(\omega)|^2 + |G(\omega)|^2 + 2|F(\omega)| \times |G(\omega)| \times \cos(\varphi_f - \varphi_g). \tag{2.67}$$

If the phase difference $(\varphi_f - \varphi_g)$ doesn't happen to be 90° (modulo 2π), the interference term does not cancel. Don't think you're on the safe side with real Fourier transforms. The phases are then 0, and the interference term reaches a maximum. The following example will illustrate this:

Example 2.13 (Overlapping lines) Let's take two spectral lines—say of shape $\frac{\sin x}{x}$— that approach each other. $H(\omega)$ simply is a linear superposition[5] of the two lines, yet not $|H(\omega)|^2$. As soon as the two lines start to overlap, there also will be an interference term. To use a concrete example, let's take the function of (2.31) and,

---

[5]i.e. addition.

Fig. 2.18 Superposition of two $\left(\frac{\sin x}{x}\right)$-functions. Power representation with interference term (*left*); power representation without interference term (*right*)

for simplicity's sake, flip the negative frequency axis to the positive axis. Then we get:

$$H_{\text{total}}(\omega) = H_1 + H_2$$
$$= T\left(\frac{\sin[(\omega - \omega_1)T/2]}{(\omega - \omega_1)T/2} + \frac{\sin[(\omega - \omega_2)T/2]}{(\omega - \omega_2)T/2}\right). \tag{2.68}$$

The phases are 0, as we have used two cosine functions $\cos \omega_1 t$ and $\cos \omega_2 t$ for input. So $|H(\omega)|^2$ becomes:

$$|H_{\text{total}}(\omega)|^2 = T^2 \left\{ \left(\frac{\sin[(\omega - \omega_1)T/2]}{(\omega - \omega_1)T/2}\right)^2 + \left(\frac{\sin[(\omega - \omega_2)T/2]}{(\omega - \omega_2)T/2}\right)^2 \right.$$
$$\left. + 2\frac{\sin[(\omega - \omega_1)T/2]}{(\omega - \omega_1)T/2} \times \frac{\sin[(\omega - \omega_2)T/2]}{(\omega - \omega_2)T/2} \right\} \tag{2.69}$$
$$= T^2 \left\{ |H_1(\omega)|^2 + H_1^*(\omega)H_2(\omega) \right.$$
$$\left. + H_1(\omega)H_2^*(\omega) + |H_2(\omega)|^2 \right\}.$$

Figure 2.18 backs up the facts: for overlapping lines, the interference term makes sure that in the power representation the lineshape is *not* the sum of the power representation of the lines. *Fix*: Show real and imaginary parts separately. If you want to keep the linear superposition (it's so useful), then you have to stay clear of the squaring!

2.5.2 Truncation Error

We now want to look at what will happen, if we truncate the function $f(t)$ somewhere—preferably where it isn't large any more—and then Fourier-transform it. Let's take a simple example:

Example 2.14 (Truncation error)

$$f(t) = \begin{cases} e^{-\lambda t} & \text{for } 0 \le t \le T \\ 0 & \text{else} \end{cases}. \tag{2.70}$$

The Fourier transform then is:

$$F(\omega) = \int_0^T e^{-\lambda t} e^{-i\omega t}\, dt = \frac{1}{-\lambda - i\omega} e^{-\lambda t - i\omega t}\Big|_0^T = \frac{1 - e^{-\lambda T - i\omega T}}{\lambda + i\omega}. \tag{2.71}$$

Compared to the untruncated exponential function, we're now saddled with the additional term $-e^{-\lambda T} e^{-i\omega T}/(\lambda + i\omega)$. For large values of T it isn't all that large, but to our grief, it oscillates. Truncating the smooth Lorentzian gave us small oscillations in return. Figure 2.19 explains that (cf. Fig. 2.7 without truncation).

The morale of the story: don't truncate if you don't have to, and most certainly neither brusquely nor brutally. How it should be done—if you've got to do it—will be explained in the next chapter.

Finally, an example how not to do it:

Example 2.15 (Exponential on pedestal) We'll once again use our truncated exponential function and put it on a pedestal, that's only nonzero between $0 \le t \le T$. Assume a height of a:

$$\begin{aligned} f(t) &= \begin{cases} e^{-\lambda t} & \text{for } 0 \le t \le T \\ 0 & \text{else} \end{cases} \quad \text{with } F(\omega) = \frac{1 - e^{-\lambda T} e^{-i\omega T}}{\lambda + i\omega}, \\ g(t) &= \begin{cases} a & \text{for } 0 \le t \le T \\ 0 & \text{else} \end{cases} \quad \text{with } G(\omega) = a\frac{1 - e^{-i\omega T}}{i\omega}. \end{aligned} \tag{2.72}$$

Here, to calculate $G(\omega)$, we've again used $F(\omega)$, with $\lambda = 0$. $|F(\omega)|^2$ we've already met in Fig. 2.19. $\text{Re}\{G(\omega)\}$ as well as $\text{Im}\{G(\omega)\}$ are shown in Fig. 2.20.

Finally, in Fig. 2.21 $|H(\omega)|^2$ is shown, decomposed into $|F(\omega)|^2$, $|G(\omega)|^2$ and the interference term.

Fig. 2.19 Fourier transform of the truncated unilateral exponential function

Fig. 2.20 Fourier transform of the pedestal

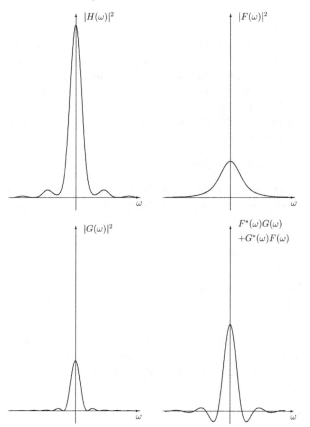

Fig. 2.21 Power representation of Fourier transform of a unilateral exponential function on a pedestal (*top left*), the unilateral exponential function (*top right*); Power representation of the Fourier transform of the pedestal (*bottom left*) and representation of the interference term (*bottom right*)

For this figure we picked the function $5e^{-5t/T} + 2$ in the interval $0 \le t \le T$. The exponential function therefore already dropped to e^{-5} at truncation, the step with $a = 2$ isn't all that high either. Therefore neither $|F(\omega)|^2$ nor $|G(\omega)|^2$ look all that terrible either, but $|H(\omega)|^2$ does. It's the interference term's fault. The truncated exponential function on the pedestal is a prototypic example for "bother" when doing Fourier transformations. As we'll see in Chap. 3, even using window functions would be of limited help. That's only the—overly popular—power representation's and interference term's fault.

Fix: Subtract the pedestal before transforming. Usually we're not interested in it anyway. For example a logarithmic representation helps, giving a straight line for the e-function, which then becomes "bent" and runs into the background. Use extrapolation to determine a. It would be best to divide by the exponential, too. You are presumably interested in (possible) small oscillations only. In case you have no data for long times, you will run into trouble. You will also get problems if you have a superposition of several exponentials such that you won't get a straight line anyhow. In such cases, I guess, you will be stumped with Fourier transformation. Here, Laplace transformation helps which we shall not treat here.

Playground

2.1 Black Magic
The Italian mathematician Maria Gaetana Agnesi—appointed in 1750 to the faculty of the University of Bologna by the Pope—constructed the following geometric locus, called "versiera":

  i. draw a circle with radius $a/2$ at $(0; a/2)$
 ii. draw a straight line parallel to the x-axis through $(0; a)$
iii. draw a straight line through the origin with a slope $\tan\theta$
 iv. the geometric locus of the "versiera" is obtained by taking the x-value from the intersection of both straight lines while the y-value is taken from the intersection of the inclined straight line with the circle.

  a. Derive the x- and y-coordinates as a function of θ, i.e. in parameterised form.
  b. Eliminate θ using the trigonometric identity $\sin^2\theta = 1/(1+\cot^2\theta)$ to arrive at $y = f(x)$, i.e. the "versiera".
  c. Calculate the Fourier transform of the "versiera".

2.2 The Phase Shift Knob
On the screen of a spectrometer you see a single spectral component with non-zero patterns for the real and imaginary parts. What shift on the time axis, expressed as a fraction of the oscillation period T, must be applied to make the imaginary part vanish? Calculate the real part which then builds up.

2.3 Pulses
Calculate the Fourier transform of:

$$f(t) = \begin{cases} \sin\omega_0 t & \text{for } -T/2 \le t \le T/2 \\ 0 & \text{else} \end{cases} \quad \text{with } \omega_0 = n\frac{2\pi}{T/2}.$$

What is $|F(\omega_0)|$, i.e. at "resonance"? Now, calculate the Fourier transform of two of such "pulses", centered at $\pm\Delta$ around $t = 0$.

2.4 Phase-Locked Pulses

Calculate the Fourier transform of:

$$f(t) = \begin{cases} \sin\omega_0 t & \text{for} \quad \begin{matrix} -\Delta - T/2 \le t \le -\Delta + T/2 \\ \text{and} \ +\Delta - T/2 \le t \le +\Delta + T/2 \end{matrix} \qquad \text{with } \omega_0 = n\dfrac{2\pi}{T/2}. \\ 0 & \text{else} \end{cases}$$

Choose Δ such that $|F(\omega)|$ is as large as possible for all frequencies ω! What is the full width at half maximum (FWHM) in this case?

Hint: Note that now the rectangular pulses "cut out" an integer number of oscillations, not necessarily starting/ending at 0, but being "phase-locked" between left and right "pulses" (Fig. 2.22).

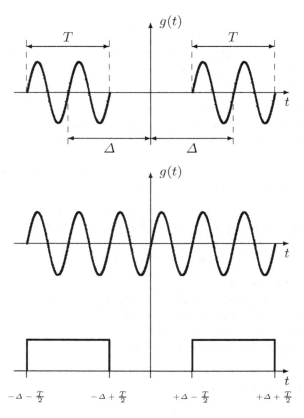

Fig. 2.22 Two pulses 2Δ apart from each other (*top*). Two "phase-locked" pulses 2Δ apart from each other (*bottom*)

2.5 Tricky Convolution

Convolve a normalised Lorentzian with another normalised Lorentzian and calculate its Fourier transform.

2.6 Even Trickier

Convolve a normalised Gaussian with another normalised Gaussian and calculate its Fourier transform.

2.7 Voigt Profile (for Gourmets only)

Calculate the Fourier transform of a normalised Lorentzian convolved with a normalised Gaussian. For the inverse transformation you need a good integration table, e.g. [8, No 3.953.2].

2.8 Derivable

What is the Fourier transform of:

$$g(t) = \begin{cases} te^{-\lambda t} & \text{for } 0 \leq t \\ 0 & \text{else} \end{cases}.$$

Is this function even, odd, or mixed?

2.9 Nothing Gets Lost

Use Parseval's theorem to derive the following integral:

$$\int_0^\infty \frac{\sin^2 a\omega}{\omega^2} d\omega = \frac{\pi}{2}a \qquad \text{with } a > 0.$$

Chapter 3
Window Functions

Abstract Various window functions are presented in their continuous formulation: rectangular, triangular, cosine, Hanning, Hamming, triplet, Gauss, Kaiser-Bessel, Blackman-Harris. A focus is on sidelobe suppression versus 3dB-bandwidth. An example is given for a test-function with a comparison of different windows. The Kaiser-Bessel window is recommended because it provides a parameter to play with and because of its monotonically decaying sidelobes.

How much fun you get out of Fourier transformations will depend very much on the proper use of window or weighting functions. F.J. Harris has compiled an excellent overview of window functions for *discrete* Fourier transformations [9]. Here we want to discuss window functions for the case of a *continuous* Fourier transformation. Porting this to the case of a discrete Fourier transformation then won't be a problem any more.

In Chap. 1 we learnt that we better stay away from transforming steps. But that's exactly what we're doing if the input signal is available for a finite time window only. Without fully realising what we were doing, we've already used the so-called rectangular window (=no weighting) on more than a few occasions. We'll discuss this window in more detail shortly.

Then we'll get into window functions where information is "switched on and off" softly. I'll promise right now that this can be good fun.

All window functions are, of course, even functions. The Fourier transforms of the window function therefore don't have an imaginary part. We require a large dynamic range so we can better compare window qualities. That's why we'll use *logarithmic* representations covering equal ranges. And that's also the reason why we can't have negative function values. To make sure they don't occur, we'll use the power representation, i.e. $|F(\omega)|^2$.

Note:

According to the Convolution Theorem, the Fourier transform of the window function represents precisely the lineshape of an undamped cosine input.

© Springer International Publishing Switzerland 2015

T. Butz, *Fourier Transformation for Pedestrians*, Undergraduate Lecture Notes in Physics, DOI 10.1007/978-3-319-16985-9_3

3.1 The Rectangular Window

$$f(t) = \begin{cases} 1 \text{ for } -T/2 \le t \le T/2 \\ 0 \text{ else} \end{cases}, \tag{3.1}$$

has the power representation of the Fourier transform:

$$|F(\omega)|^2 = T^2 \left(\frac{\sin(\omega T/2)}{\omega T/2} \right)^2. \tag{3.2}$$

The rectangular window and this function are shown in Fig. 3.1. The first sidelobe is negative and all subsequent sidelobes alternate in sign.

3.1.1 Zeros

Where are the zeros of this function? We'll find them at $\omega T/2 = \pm l\pi$ with $l = 1, 2, 3, \ldots$ and without 0! The zeros are equidistant, the zero at $l = 0$ in the numerator gets "plugged" by a zero at $l = 0$ in the denominator.[1]

3.1.2 Intensity at the Central Peak

Now we want to find out how much intensity is at the central peak, and how much gets lost in the sidebands (sidelobes). To get there, we need the first zero at $\omega T/2 = \pm\pi$ or $\omega = \pm 2\pi/T$ and:

$$\int_{-2\pi/T}^{+2\pi/T} T^2 \left(\frac{\sin(\omega T/2)}{\omega T/2} \right)^2 d\omega = T^2 \frac{2}{T} 2 \int_0^\pi \frac{\sin^2 x}{x^2} dx = 4T \, \mathrm{Si}(2\pi) \tag{3.3}$$

$$\text{where } \omega T/2 = x.$$

Here $\mathrm{Si}(x)$ stands for the sine integral:

$$\int_0^x \frac{\sin y}{y} dy. \tag{3.4}$$

---

[1] Use l'Hospital's rule for $\lim_{x \to 0} \frac{\sin x}{x}$.

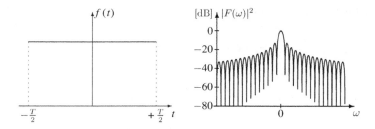

Fig. 3.1 Rectangular window function and its Fourier transform in power representation (the unit dB, "decibel", will be explained in Sect. 3.1.3)

The last equal sign may be proved as follows. We start out with:

$$\int_0^\pi \frac{\sin^2 x}{x^2} dx$$

and integrate partially with $u = \sin^2 x$ and $v = -\frac{1}{x}$:

$$\int_0^\pi \frac{\sin^2 x}{x^2} dx = \frac{\sin^2 x}{x}\Big|_0^\pi + \int_0^\pi \frac{2\sin x \cos x}{x} dx$$

$$= 2\int_0^\pi \frac{\sin 2x}{2x} dx = \text{Si}(2\pi) \tag{3.5}$$

$$\text{with } 2x = y.$$

Using Parseval's theorem we get the total intensity:

$$\int_{-\infty}^{+\infty} T^2 \left(\frac{\sin(\omega T/2)}{\omega T/2}\right)^2 d\omega = 2\pi \int_{-T/2}^{+T/2} 1^2 dt = 2\pi T. \tag{3.6}$$

The ratio of the intensity at the central peak to the total intensity therefore is:

$$\frac{4T\,\text{Si}(2\pi)}{2\pi T} = \frac{2}{\pi}\text{Si}(2\pi) = 0.903.$$

This means that $\approx 90\%$ of the intensity is in the central peak, whereas some 10% are "wasted" in sidelobes.

3.1.3 Sidelobe Suppression

Now let's determine the height of the first sidelobe. To get there, we need:

$$\frac{\mathrm{d}|F(\omega)|^2}{\mathrm{d}\omega} = 0 \quad \text{or also} \quad \frac{\mathrm{d}F(\omega)}{\mathrm{d}\omega} = 0 \tag{3.7}$$

and that's the case when:

$$\frac{\mathrm{d}}{\mathrm{d}x}\frac{\sin x}{x} = 0 = \frac{x\cos x - \sin x}{x^2} \quad \text{with } x = \omega T/2 \text{ or } x = \tan x.$$

Solving this transcendental equation (for example graphically or by trial and error) gives us the smallest possible solution $x = 4.4934$ or $\omega = 8.9868/T$. Inserting that in $|F(\omega)|^2$ results in:

$$\left|F\left(\tfrac{8.9868}{T}\right)\right|^2 = T^2 \times 0.04719. \tag{3.8}$$

For $\omega = 0$ we get $|F(0)|^2 = T^2$, the ratio of the first sidelobe's height to the central peak's height therefore is 0.04719. It's customary to express ratios between two values spanning several orders of magnitude in decibels (short: dB). The definition of the decibel is:

$$\mathrm{dB} = 10\log_{10} x \tag{3.9}$$

Quite regularly people forget to mention *what* the ratio's based on, which can cause confusion. We're talking about intensity-ratios, (viz. $F^2(\omega)$). If we're referring to amplitude-ratios, (viz. $F(\omega)$), this would make precisely a factor of two in logarithmic representation! Here we have a sidelobe suppression (first sidelobe) of:

$$10\log_{10} 0.04719 = -13.2\,\mathrm{dB}. \tag{3.10}$$

3.1.4 3 dB-Bandwidth

As the $10\log_{10}(1/2) = -3.0103 \approx -3$, the 3 dB bandwidth tells us where the central peak has dropped to half its height. This is easily calculated as follows:

$$T^2\left(\frac{\sin(\omega T/2)}{\omega T/2}\right)^2 = \frac{1}{2}T^2.$$

Using $x = \omega T/2$ we have:

$$\sin^2 x = \frac{1}{2}x^2 \quad \text{or} \quad \sin x = \frac{1}{\sqrt{2}}x. \tag{3.11}$$

This transcendental equation has the following solution:

$$x = 1.3915, \qquad \text{thus} \qquad \omega_{3\text{dB}} = 2.783/T.$$

This gives us the total width ($\pm \omega_{3\text{dB}}$):

$$\Delta \omega = \frac{5.566}{T}. \tag{3.12}$$

This is the slimmest central peak we can get using Fourier transformation. Any other window function will lead to larger 3 dB-bandwidths. Admittedly, it's more than nasty to stick more than $\approx 10\%$ of the information into the sidelobes. If we have, apart from the prominent spectral component, another spectral component, with—say—an approx. 10 dB smaller intensity, this component will be completely smothered by the main component's sidelobes. If we're lucky, it will sit on the first sidelobe and will be visible; if we're out of luck, it will fall into the gap (the zero) between central peak and first sidelobe and will get swallowed. So it pays to get rid of these sidelobes.

Warning: This 3 dB-bandwidth is valid for $|F(\omega)|^2$ and not for $F(\omega)$! Since one often uses $|F(\omega)|$ or the cosine-/sine-transformation (cf. Sect. 4.5) one wants the 3 dB-bandwidth thereof, which corresponds to the 6 dB-bandwidth of $|F(\omega)|^2$. Unfortunately, you cannot simply multiply the 3 dB-bandwidth of $|F(\omega)|^2$ by $\sqrt{2}$, you have to solve a new transcendental equation. However, it's still good as a first guess because you merely interpolate linearly between the point of 3 dB-bandwidth and the point of the 6 dB-bandwidth. You'd overestimate the width by less than 5 %.

3.1.5 Asymptotic Behaviour of Sidelobes

The sidelobes' envelope results in the heights decreasing by 6 dB per octave (that's a factor of 2 as far as the frequency is concerned). This result is easily derived from (1.62). The unit step leads to oscillations which decay as $\frac{1}{k}$, i.e. in the continuous case as $\frac{1}{\omega}$. This corresponds to a decay of 3 dB per octave. Now we are dealing with squared magnitudes, hence, we have a decay of $\frac{1}{\omega^2}$. This corresponds to a decay of 6 dB per octave. This is of fundamental importance: a discontinuity in the function yields -6 dB/octave, a discontinuity in the derivative (hence, a kink in the function) yields -12 dB/octave and so forth. This is immediately clear considering that the derivative of the triangular function yields the step function. The derivative of $\frac{1}{\omega}$ yields $\frac{1}{\omega^2}$ (apart from the sign), i.e. a factor of 2 in the sidelobe suppression. You remember the $\left(\frac{1}{k^2}\right)$-dependence of the Fourier coefficients of the triangular function? The "smoother" the window function starts out, the better the sidelobes' asymptotic behaviour will get. But this comes at a price, namely a worse 3 dB-bandwidth.

3.2 The Triangular Window (Fejer Window)

The first real weighting function is the triangular window:

$$f(t) = \begin{cases} 1 + 2t/T & \text{for } -T/2 \le t \le 0 \\ 1 - 2t/T & \text{for } 0 \le t \le T/2 \\ 0 & \text{else} \end{cases} , \qquad (3.13)$$

$$F(\omega) = \frac{T}{2} \left(\frac{\sin(\omega T/4)}{\omega T/4} \right)^2 . \qquad (3.14)$$

We won't have to rack our brains! This is the autocorrelation function of the triangular function (cf. Sect. 2.3.1, Fig. 2.12). The only difference is the interval's width: whereas the autocorrelation function of the rectangular function over the interval $-T/2 \le t \le T/2$ has a width of $-T \le t \le T$, in (3.13) we only have the usual interval $-T/2 \le t \le T/2$.

The 1/4 is due to the interval, the square due to the autocorrelation. All other properties are obvious straight away. The triangular window and the square of this function are shown in Fig. 3.2.

The zeros are twice as far apart as in the case of the rectangular function:

$$\frac{\omega T}{4} = \pi l \quad \text{or} \quad \omega = \frac{4\pi l}{T} \quad l = 1, 2, 3, \dots . \qquad (3.15)$$

The intensity at the central peak is 99.7 %.

The height of the first sidelobe is suppressed by $2 \times (-13.2 \text{ dB}) \approx -26.5 \text{ dB}$ (No wonder, if we skip every other zero!).

Fig. 3.2 Triangular window and power representation of the Fourier transform

The 3 dB-bandwidth is calculated as follows:

$$\sin\frac{\omega T}{4} = \frac{1}{\sqrt[4]{2}}\frac{\omega T}{4} \quad \text{to} \quad \Delta\omega = \frac{8.016}{T} \text{ (full width)}, \tag{3.16}$$

that's some 1.44 times wider than in the case of the rectangular window.

The asymptotic behaviour of the sidelobes is -12 dB/octave.

3.3 The Cosine Window

The triangular window had a kink when switching on, another kink at the maximum ($t = 0$) and another one when switching off. The cosine window avoids the kink at $t = 0$:

$$f(t) = \begin{cases} \cos\frac{\pi t}{T} & \text{for } -T/2 \leq t \leq T/2 \\ 0 & \text{else} \end{cases}. \tag{3.17}$$

The Fourier transform of this function is:

$$F(\omega) = T\cos\frac{\omega T}{2} \times \left(\frac{1}{\pi - \omega T} + \frac{1}{\pi + \omega T}\right). \tag{3.18}$$

The functions $f(t)$ and $|F(\omega)|^2$ are shown in Fig. 3.3.

At position $\omega = 0$ we get:

$$F(0) = \frac{2T}{\pi}.$$

For $\omega T \to \pm\pi$ we get expressions of type "0 : 0", which we calculate using l'Hospital's rule.

Surprise surprise: The zero at $\omega T = \pm\pi$ was "plugged" by the expression in brackets in (3.18), i.e. $F(\omega)$ there will stay finite and continuous. Apart from that, the following applies:

Fig. 3.3 Cosine window and power representation of the Fourier transform

The zeros are at:

$$\frac{\omega T}{2} = \frac{(2l+1)\pi}{2}, \qquad \omega = \frac{(2l+1)\pi}{T}, \qquad l = 1, 2, 3, \ldots, \tag{3.19}$$

i.e. within the same distance as in the case of the rectangular window.

Here it's not worth shedding tears for a lack of intensity at the central peak any more. For all practical purposes it is $\approx 100\%$. We should, however, have another look at the sidelobes because of the minorities, viz. the chance of detecting additional weak signals.

The suppression of the first sidelobe may be calculated as follows:

$$\tan \frac{x}{2} = \frac{4x}{\pi^2 - x^2} \qquad \text{with the solution } x \approx 11.87. \tag{3.20}$$

This results in a sidelobe suppression of $-23\,\text{dB}$.

The 3 dB-bandwidth amounts to:

$$\Delta\omega = \frac{7.47}{T}, \tag{3.21}$$

a remarkable result. This is the first time we got, through the use of a somewhat more intelligent "window", a sidelobe suppression of $-23\,\text{dB}$—not a lot worse than the $-26.5\,\text{dB}$ of the triangular window—and we get a better 3 dB-bandwidth compared to $\Delta\omega = 8.016/T$ for the triangular window. So it does pay to think about better window functions. The asymptotic decay of the sidelobes is $-12\,\text{dB/octave}$, as was the case for the triangular function.

3.4 The $\cos^2$-Window (Hanning)

The scientist Julius von Hann thought that eliminating the kinks at $\pm T/2$ would be beneficial and proposed the $\cos^2$-window (in the US, this soon was called "Hanning"):

$$f(t) = \begin{cases} \cos^2 \frac{\pi t}{T} & \text{for } -T/2 \le t \le T/2 \\ 0 & \text{else} \end{cases}. \tag{3.22}$$

The corresponding Fourier transform is:

$$F(\omega) = \frac{T}{4} \sin \frac{\omega T}{2} \times \left(\frac{1}{\pi - \omega T/2} + \frac{2}{\omega T/2} - \frac{1}{\pi + \omega T/2} \right). \tag{3.23}$$

The functions $f(t)$ and $|F(\omega)|^2$ are shown in Fig. 3.4.

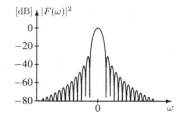

Fig. 3.4 Hanning window and power representation of the Fourier transform

The zero at $\omega = 0$ has been "plugged" because of $\sin(\omega T/2)/(\omega T/2) \to 1$ and the zeros at $\omega = \pm 2\pi/T$ get "plugged" for the same reason. The example of the cosine window is becoming popular!

The zeros are at:

$$\omega = \pm\frac{2l\pi}{T}, \qquad l = 2, 3, \ldots . \tag{3.24}$$

Intensity at the central peak $\approx 100\%$.

The suppression of the first sidelobe is $-32\,\text{dB}$.

The 3 dB-bandwidth is:

$$\Delta\omega = \frac{9.06}{T}. \tag{3.25}$$

The sidelobes' asymptotic decay is $-18\,\text{dB/octave}$.

So we get a considerable sidelobe suppression, admittedly to the detriment of the 3 dB-bandwidth.

Some experts recommend to go for higher-powered cosine functions in the first place. This would "plug" more and more zeros near the central peak, and there will be gains both as far as sidelobe suppression as well as asymptotic behaviour are concerned, though, of course, the 3 dB-bandwidth will get bigger and bigger. So for the cos³-window we get:

$$\Delta\omega = \frac{10.4}{T} \tag{3.26}$$

and for the cos⁴-window:

$$\Delta\omega = \frac{11.66}{T}. \tag{3.27}$$

As we'll see shortly, there are more intelligent solutions to this problem.

3.5 The Hamming Window

Mr Julius von Hann didn't have a clue that he—sorry: his window function—would be put on a pedestal in order to get an even better window, and to add insult to injury, his name would get mangled to "Hamming" to boot.[2]

$$f(t) = \begin{cases} a + (1 - a)\cos^2 \frac{\pi t}{T} & \text{for } -T/2 \le t \le T/2 \\ 0 & \text{else} \end{cases} . \qquad (3.28)$$

The Fourier transform is:

$$F(\omega) = \frac{T}{4}\sin\frac{\omega T}{2} \times \left(\frac{1-a}{\pi - \omega T/2} + \frac{2(1+a)}{\omega T/2} - \frac{1-a}{\pi + \omega T/2} \right). \qquad (3.29)$$

How come there's a "pedestal"? Didn't we realise a few moments ago that any discontinuity at the interval boundaries is "bad"? Just like a smidgen of arsenic may work wonders, here a "tiny wee pedestal" can be helpful. Indeed, using parameter a we're able to play the sidelobes a bit. A value of $a \approx 0.1$ proves to be good. The plugging of the zeros hasn't changed, as (3.29) shows. Though now, however, the Fourier transform of the "pedestal" has saddled us with the term:

$$\frac{T}{2}a\frac{\sin(\omega T/2)}{\omega T/2}$$

that now gets added to the sidelobes of the Hamming window. A squaring of $F(\omega)$ is not essential here. This on the one hand will provide interference terms of the Hamming window's Fourier transform, but on the other hand, the same is true for $F(\omega)$; here all we get are positive and negative sidelobes. The absolute values of the sidelobes' heights don't change. The Hamming window with $a = 0.15$ and the respective $F^2(\omega)$ are shown in Fig. 3.5. The first sidelobes are slightly smaller than the second ones! Here we have the same zeros as (this is done by the $\sin\frac{\omega T}{2}$, provided the denominators don't prevent it). For the optimal parameter $a = 0.08$ the sidelobe suppression is $-43\,\text{dB}$, the 3 dB-bandwidth is only $\Delta\omega = 8.17/T$. The asymptotic behaviour, naturally, got worse. Far from the central peak, it's down to as little as $-6\,\text{dB}$ per octave. That's what happens when you choose a small step!

Therefore the new strategy is: rather a somewhat worse asymptotic behaviour, if only we manage to get a high sidelobe suppression and, at the same time, a decrease in 3 dB-bandwidth deterioration that's as small as possible. How far one can go is illustrated by the following example. Plant at the interval ends little "flagpoles", i.e. infinitely sharp cusps with small height. This is, of course, most easily done in the discrete Fourier transformation. There, the "flagpole" is just a channel wide. Of

---

[2]No kidding, Mr R.W. Hamming apparently did discover this window, and the von Hann window got mangled later on.

Fig. 3.5 Hamming window and power representation of Fourier transform

course, we get no asymptotic roll-off of the sidelobes at all. The Fourier transform of a δ-function is a constant! However, we get the narrowest 3 dB-bandwidth possible. Such a window is called Dolph–Chebychev window, however, we won't discuss it any further here. Because it emphasizes the data at the interval boundaries it should not be used in cases where those data are inaccurate or corrupted.

Before we get into more and better window functions, let's look, just for curiosity's sake, at a window that creates no sidelobes at all.

3.6 The Triplet Window

The previous really set us up, so let's try the following:

$$f(t) = \begin{cases} e^{-\lambda|t|} \cos^2 \frac{\pi t}{T} & \text{for } -T/2 \leq t \leq T/2 \\ 0 & \text{else} \end{cases}. \tag{3.30}$$

Deducing the expression for $F(\omega)$ is trivial, yet too lengthy (and too unimportant) to be dealt with here.

The expression for $F(\omega)$—if we do deduct it—stands out, as it features oscillating terms (sine, cosine) though there are no more zeros. If only the λ is big

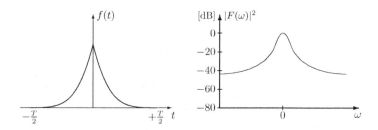

Fig. 3.6 Triplet window and power representation of the Fourier transform

enough, then there won't even be local minima or maxima any more, and $F(\omega)$ decays monotonically. In the case of optimum λ we can achieve an asymptotic behaviour of -12 dB/octave with a 3 dB-bandwidth of $\Delta\omega = 9.7/T$ (cf. Fig. 3.6).

Therefore it wasn't such a bad idea to re-introduce a spike at $t = 0$. However, there are better window functions.

3.7 The Gauss Window

A pretty obvious window function is the Gauss function. That we have to truncate it somewhere, resulting in a small step, doesn't worry us any more, if we look back on our experience with the Hamming window.

$$f(t) = \begin{cases} \exp\left(-\frac{1}{2}\frac{t^2}{\sigma^2}\right) & \text{for } -T/2 \le t \le +T/2 \\ 0 & \text{else} \end{cases}. \tag{3.31}$$

The Fourier transform reads:

$$F(\omega) = \sigma\sqrt{\frac{\pi}{2}}e^{-\frac{\sigma^2\omega^2}{2}}\left(\text{erf}\left(-\frac{i\sigma\omega}{\sqrt{2}} + \frac{T}{2\sqrt{2}\sigma}\right) + \text{erf}\left(+\frac{i\sigma\omega}{\sqrt{2}} + \frac{T}{2\sqrt{2}\sigma}\right)\right). \tag{3.32}$$

As the error function occurs with complex arguments, though together with the conjugate complex argument, $F(\omega)$ is real. The function $f(t)$ with $\sigma = 2$ in units of $T/2$ and $|F(\omega)|^2$ is shown in Fig. 3.7. A disadvantage of the Gauss window is that its sidelobes do not decay monotonically for all values of σ.

A Gauss function being Fourier-transformed will result in another Gauss function, yet only when there was no truncation! If σ is sufficiently big, the sidelobes will disappear: the oscillations "creep up" the Gauss function's flank. Shortly before this happens, we get a 3 dB-bandwidth of $\Delta\omega = 9.06/T$, -64 dB sidelobe suppression and -26 dB per octave asymptotic behaviour near the central peak. Not bad, but we can do better.

Fig. 3.7   Gauss window and power representation of the Fourier transform

3.8 The Kaiser–Bessel Window

The Kaiser–Bessel window is a very useful window and can be applied to various situations:

$$f(t) = \begin{cases} \dfrac{I_0\left(\beta\sqrt{1-(2t/T)^2}\right)}{I_0(\beta)} & \text{for } -T/2 \leq t \leq T/2 \\ 0 & \text{else} \end{cases} \tag{3.33}$$

Here β is a parameter that may be chosen at will. The Fourier transform is:

$$F(\omega) = \begin{cases} \dfrac{2T}{I_0(\beta)}\,\dfrac{\sinh\left(\sqrt{\beta^2 - \frac{\omega^2 T^2}{4}}\right)}{\sqrt{\beta^2 - \frac{\omega^2 T^2}{4}}} & \text{for } \beta \geq \left|\frac{\omega T}{2}\right| \\[4mm] \dfrac{2T}{I_0(\beta)}\,\dfrac{\sin\left(\sqrt{\frac{\omega^2 T^2}{4} - \beta^2}\right)}{\sqrt{\frac{\omega^2 T^2}{4} - \beta^2}} & \text{for } \beta \leq \left|\frac{\omega T}{2}\right| \end{cases} \tag{3.34}$$

$I_0(x)$ is the modified Bessel function. A simple algorithm [10, Equations 9.8.1, 9.8.2] for the calculation of $I_0(x)$ follows:

$$I_0(x) = 1 + 3.5156229t^2 + 3.0899424t^4 + 1.2067492t^6$$
$$+ 0.2659732t^8 + 0.0360768t^{10} + 0.0045813t^{12} + \epsilon,$$
$$|\epsilon| < 1.6 \times 10^{-7}$$
$$\text{with } t = x/3.75, \text{ for the interval } -3.75 \leq x \leq 3.75,$$

and:

$$x^{1/2}e^{-x}I_0(x) = 0.39894228 + 0.01328592t^{-1} + 0.00225319t^{-2}$$
$$- 0.00157565t^{-3} + 0.00916281t^{-4} - 0.02057706t^{-5}$$
$$+ 0.02635537t^{-6} - 0.01647633t^{-7} + 0.00392377t^{-8} + \epsilon,$$
$$|\epsilon| < 1.9 \times 10^{-7}$$
$$\text{with } t = x/3.75, \text{ for the interval } 3.75 \leq x < \infty.$$

The zeros are at $\omega^2 T^2/4 = l^2\pi^2 + \beta^2$, $l = 1, 2, 3, \ldots$, and they're not equidistant. For $\beta = 0$ we get the rectangular window, values up to $\beta = 9$ are recommended. Figure 3.8 shows $f(t)$ and $|F(\omega)|^2$ for various values of β.

The sidelobe suppression as well as the 3 dB-bandwidth as a function of β are shown in Fig. 3.9. Using this window function we get for $\beta = 9$ -70 dB sidelobe suppression with $\Delta\omega = 11/T$ and -38.5 dB/octave asymptotic behaviour near the central peak. In every respect, the Kaiser–Bessel windows is superior to the Gauss window.

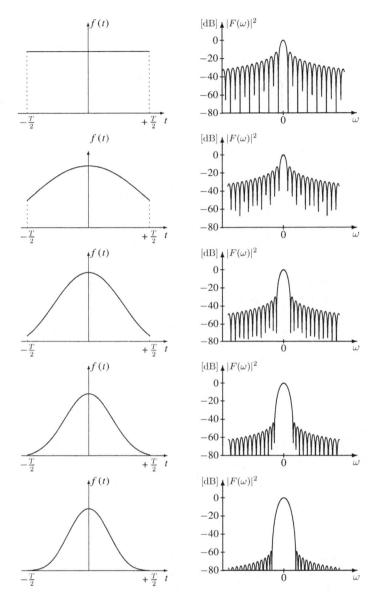

Fig. 3.8 Kaiser–Bessel window for $\beta = 0, 2, 4, 6, 8$ (*left*) and the respective power representation of the Fourier transform (*right*)

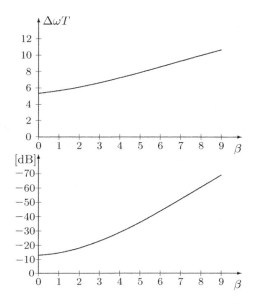

Fig. 3.9 Sidelobe suppression and 3 dB-bandwidth for Kaiser–Bessel parameter $\beta = 0 - 9$

3.9 The Blackman–Harris Window

To those of you who don't want flexibility and want to work with a fixed good sidelobe suppression, I recommend the following two very efficient windows which are due to Blackman and Harris. They have the charm to be simple: they consist of a sum of four cosine terms as follows:

$$f(t) = \begin{cases} \sum_{n=0}^{3} a_n \cos \frac{2\pi n t}{T} & \text{for } -T/2 \leq t \leq T/2 \\ 0 & \text{else} \end{cases}. \tag{3.35}$$

Please note that we have a constant, a cosine term with a full period, as well as further terms with two and three full periods, contrary to the Sect. 3.3. Here, the coefficients have the following values:

	for -74 dB	for -92 dB
a_0	0.40217	0.35875
a_1	0.49703	0.48829
a_2	0.09392	0.14128
a_3	0.00183	0.01168 .

(3.36)

In the original publication of Harris [9] the sum the coefficients of the -74 dB window is smaller than 1 by 0.00505. There must be a misprint. With the coefficients listed above the "sidelobe" suppression is significantly worse than -74 dB. If you take $a_1 = 0.49708$ and $a_2 = 0.09892$ (or $a_2 = 0.09892$ and $a_3 = 0.00188$) you obtain -74 dB. In this case the sum of the coefficients yields 1. Maybe, in 1978 there were problems during the typing of the manuscript: the character 8 was read as 3? The second choice is that given in [11].

Surely, you have noted that the coefficients add up to 1; at the interval ends the terms with a_0 and a_2 are positive, whereas the terms with a_1 and a_3 are negative. The sum of the even coefficients minus the sum of the odd coefficients yields practically 0, i.e. there is a rather "soft" turning on with a tiny step only.

The Fourier transform of this window reads:

$$F(\omega) = T \sin \frac{\omega T}{2} \sum_{n=0}^{3} a_n (-1)^n \left(\frac{1}{2n\pi + \omega T} - \frac{1}{2n\pi - \omega T} \right). \tag{3.37}$$

Don't worry, the zeros in the denominator are just "healed" by the zeros of the sine. The zeros of the Fourier transform are given by $\sin \frac{\omega T}{2} = 0$, i.e. they are the same as for the Hanning window. The 3 dB-bandwidth is $\Delta \omega = 10.93/T$ and $11.94/T$ for the -74 dB- and the -92 dB-window, respectively; excellent performance for such simple windows. I guess, the series expansion of the modified Bessel function $I_0(x)$ for the appropriate values of β yields pretty much the coefficients of the Blackman–Harris windows. Because these Blackman–Harris windows differ only very little from the Kaiser–Bessel windows with $\beta \approx 9$ and $\beta \approx 11.5$, respectively, (these are the values for comparable sidelobe suppression), I do without figures. However, the Blackman–Harris window with -92 dB would have no more visible "feetlets" in Fig. 3.10 which displays to -80 dB only.

3.10 Overview over Window Functions

In order to fill this chapter with life we give a simple example. Given is the following function:

$$\begin{aligned} f(t) = {}&\cos \omega t + 10^{-2} \cos 1.15\omega t + 10^{-3} \cos 1.25\omega t \\ &+ 10^{-3} \cos 2\omega t + 10^{-4} \cos 2.75\omega t + 10^{-5} \cos 3\omega t. \end{aligned} \tag{3.38}$$

Apart from the dominant frequency ω there are two satellites at 1.15 and 1.25 times ω, two harmonics—radio frequency technicians say first and second harmonic— at 2ω and 3ω as well as another frequency at 2.75ω. Let's Fourier-transform this function. Please keep in mind that we shall look at the power spectra right now, i.e. the amplitudes squared! Hence, the signs of the amplitudes play no role. Apart from the dominant frequency, which we will quote with 0 dB intensity, we expect further spectral components with intensities of -40, -60, -80 and -100 dB.

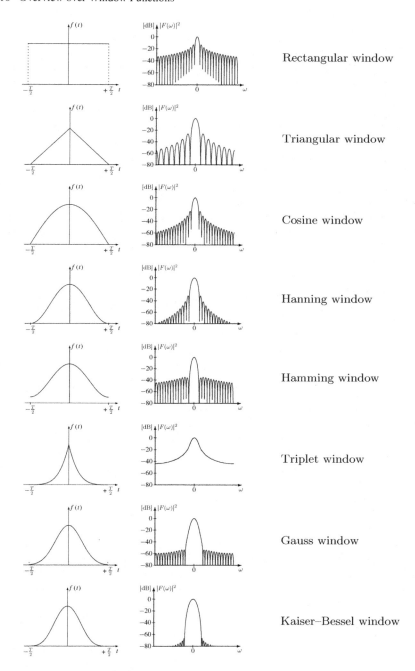

Fig. 3.10 "Overview" of the window functions

Fig. 3.11 Test function from (3.38) analysed with different window functions

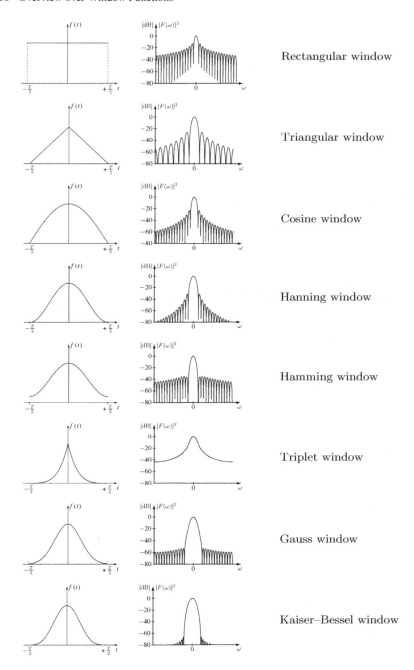

Fig. 3.10 "Overview" of the window functions

Fig. 3.11 Test function from (3.38) analysed with different window functions

Figure 3.11 shows what you get using different window functions. For the purists: of course, we have used the discrete Fourier transform to be dealt with in the next chapter, but show line-plots (we have used 128 data points, zero-padded the data, mirrored and used a total of 4096 input data; now you can repeat it yourself!).[3]

The two satellites close to the dominant frequency cause the biggest problems. On the one hand we require a window function with a good sidelobe suppression in order to be able to see the signals with intensities of −40 and −60 dB. The rectangular window does'nt achieve that! You only see the dominant frequency, all the rest is "drowned". In addition, we require a small 3 dB-bandwidth in order to resolve the frequency which is 15 % higher. This is pretty well accomplished using the Hanning- and above all the Hamming-window (Parameter 0.08). However, the Hamming window is unable to detect the higher spectral components which still have lower intensities. This is a consequence of the poor asymptotic behaviour. We are no better off with the component which is 25 % higher because it has −60 dB intensity only. Here, the Blackman–Harris window with −74 dB is just able to do so. It is easy to detect the other three, still higher spectral components, regardless of their low intensities, because they are far away from the dominant frequency if only the sidelobes in this spectral range are not "drowning" them. Interestingly enough, window functions with poor sidelobe suppression but good asymptotic behaviour like the Hanning window are doing the job, as do window functions with good sidelobe suppression and poor asymptotic behaviour like the Kaiser–Bessel window. The Kaiser–Bessel window with the parameter $\beta = 12$ is an example (the Blackman–Harris window with −92 dB sidelobe suppression is nearly as good). The disadvantage: the small satellites at 1.15- and 1.25-fold frequency show up as shoulders only. You see that we should use different window functions for different demands. There is no multi-purpose beast providing eggs, wool, milk and bacon! However, there are window functions which you can simply forget. Since the Kaiser–Bessel window has an adjustable parameter β, I recommend to try out various values of β. An important property of this window is its monotonic decay of the sidelobes.

What can we do if we need a lot more sidelobe suppression than −100 dB? Take the Kaiser–Bessel window with a very large parameter β; you easily get much better sidelobe suppression, of course with increasingly larger 3 dB-bandwidth! There is no escape from this "double mill"! However, despite the joy about "intelligent" window functions you should not forget that first you should obtain data which contain so little noise that they allow the mere detection of −100 dB-signals.

---

[3] I must confess, all the previous figures of the Fourier-transformed window functions were prepared by discrete Fourier transformation and there are subtle differences compared to the analytical formulas like those shown in Fig. 4.19.

3.11 Windowing or Convolution?

In principle, we have two possibilities to use window functions:

i. either you weight, i.e. you multiply the input by the window function and subsequently Fourier-transform, or
ii. you Fourier-transform the input and convolve the result with the Fourier transform of the window function.

According to the Convolution Theorem (2.42) we get the same result. What are the pros and cons of both procedures? There is no easy answer to this question. What helps in arguing is thinking in discrete data. Take, e.g. the Kaiser–Bessel window. Let's start with a reasonable value for the parameter β, based on considerations of the trade-off between 3 dB-bandwidth (i.e. resolution) and sidelobe suppression. In the case of windowing we have to multiply our input data, say N real or complex numbers, by the window function which we have to calculate at N points. After that we Fourier-transform. Should it turn out that we actually should require a better sidelobe suppression and could tolerate a worse resolution—or vice versa—we would have to go back to the original data, window them again and Fourier-transform again.

The situation is different for the case of convolution: we Fourier-transform without any bias concerning the eventually required sidelobe suppression and subsequently convolve the Fourier data (again N numbers, however in general complex!) with the Fourier-transformed window function, which we have to calculate for a sufficient number of points. What is a sufficient number? Of course, we drop the sidelobes for the convolution and only take the central peak! This should be calculated at least for 5 points, better more. The convolution then actually consists of 5 (or more) multiplications and a summation for each Fourier coefficient. This appears to be more work; however, it has the advantage that a further convolution with another, say broader Fourier-transformed window function, would not require to carry out a new Fourier transformation. Of course, this procedure is but an approximation because of the truncation of the sidelobes. If we included all data of the Fourier-transformed window function including the sidelobes, we had to carry out N (complex) multiplications and a summation per point, already quite a lot of computational effort, yet still less than a new Fourier transformation. This could be relevant for large arrays, especially in two or three dimensions like in image processing and tomography.

What happens at the edges when carrying out a convolution? We shall see in the following chapter that we shall continue periodically beyond the interval. This gives us the following idea: let's take the Blackman–Harris window and continue periodically; the corresponding Fourier transform consists of a sum of four δ-functions, in the discrete world we have exactly four channels which are non-zero. Where remained the sidelobes? You shall see in a minute that in this case the points (by the way equidistant) coincide with the zeros of the Fourier-transformed window function, except at 0! Hence, we have to carry out a convolution with just four points only, a rather fast procedure! That's why the Blackman–Harris window is called a 4-point-window. So after all, convolution is better? Here comes a deep sigh: there are so many good reasons to get rid of the periodic continuation as much as possible by zero-padding the input data (cf. Sect. 4.6), thus our neat 4-point-idea melts away like snow in springtime sun. The decision is yours whether you prefer to weight or to convolve and depends on the concrete case. Now it's high time to start with the discrete Fourier transformation!

Playground

3.1 Squared
Calculate the 3 dB-bandwidth of $F(\omega)$ for the rectangular window. Compare this with the 3 dB-bandwidth $F^2(\omega)$.

3.2 Let's Gibbs Again
What is the asymptotic behaviour of the Gauss window far away from the central peak?

3.3 Expander
The series expansion of the modified Bessel function of zeroth order is:

$$I_0(x) = \sum_{k=0}^{\infty} \frac{(x^2/4)^k}{(k!)^2},$$

where $k! = 1 \times 2 \times 3 \times \cdots \times k$ denotes the factorial. The series expansion for the cosine reads:

$$\cos(x) = \sum_{k=0}^{\infty} (-1)^k \frac{x^{2k}}{(2k)!}.$$

Calculate the first ten terms in the series expression of the Blackman–Harris window with -74 dB sidelobe suppression and the Kaiser–Bessel window with $\beta = 9$ and compare the results.

Hint: Instead of pen and paper better use your PC!

3.4 Minorities
In a spectrum analyser you detect a signal at $\omega = 500$ Mrad/s in the $|F(\omega)|^2$-mode with an instrumental full width at half maximum (FWHM) of 50 Mrad/s with a rectangular window.

a. What sampling period T did you choose?
b. What window function could you use if you were hunting a "minority" signal which you suspect to be 20 % higher in frequency and 50 dB lower than the main signal. Look at the figures in this chapter, don't calculate too much.

Chapter 4
Discrete Fourier Transformation

Abstract This chapter deals with the discrete Fourier transformation. Here, a periodic series in the time domain is mapped onto a periodic series in the frequency domain. Definitions of the discrete Fourier transformation and its inverse are given. Linearity, convolution, cross-correlation, and autocorrelation are treated as well as Parseval's theorem. The sampling theorem is illustrated with a simple example. Data mirroring, cosine- and sine-transformations, as well as zero-padding are discussed. It concludes with the Fast Fourier Transformation algorithm by Cooley and Tukey.

Mapping of a *Periodic* Series $\{f_k\}$ to the Fourier-Transformed Series $\{F_j\}$

4.1 Discrete Fourier Transformation

Often we don't know a function's continuous "behaviour" over time, but only what happens at N discrete times:

$$t_k = k\,\Delta t, \qquad k = 0, 1, \ldots, N-1.$$

In other words: we've taken our "pick", that's "samples" $f(t_k) = f_k$ at certain points in time t_k. Any digital data-recording uses this technique. So the data set consists of a series $\{f_k\}$. Outside the sampled interval $T = N\,\Delta t$ we don't know anything about the function. The discrete Fourier transformation automatically assumes, that $\{f_k\}$ will continue periodically outside the interval's range. At first glance this limitation appears to be very annoying: maybe $f(t)$ isn't periodic at all, and even if $f(t)$ were periodic, there's a chance that our interval happens to truncate at the wrong time (meaning: not after an integer number of periods). How this problem can be alleviated or practically eliminated will be shown in Sect. 4.6. To make life easier, we'll also take for granted that N is a power of 2. We'll have to assume the latter anyway for the Fast Fourier Transformation (FFT) which we'll cover in Sect. 4.7. Using the "trick" from Sect. 4.6, however, this limitation will become completely irrelevant.

© Springer International Publishing Switzerland 2015   93
T. Butz, *Fourier Transformation for Pedestrians*, Undergraduate Lecture
Notes in Physics, DOI 10.1007/978-3-319-16985-9_4

4.1.1 Even and Odd Series and Wrap-Around

A series is called even if the following is true for all k:

$$f_{-k} = f_k. \tag{4.1}$$

A series is called odd if the following is true for all k:

$$f_{-k} = -f_k. \tag{4.2}$$

(Here $f_0 = 0$ is compulsory!). Any series can be broken up into an even and an odd series. But what about negative indices? We'll extend the series periodically:

$$f_{-k} = f_{N-k}. \tag{4.3}$$

This allows us, by adding N, to shift the negative indices to the right end of the interval, or using another word, "wrap them around", as shown in Fig. 4.1.

Please make sure f_0 doesn't get wrapped, something that often is done by mistake. The periodicity with period N, which we *always* assume as given for the discrete Fourier transformation, requires $f_N = f_0$. In the second example—the one with the mistake—we would get f_0 twice next to each other (and apart from that, we would have overwritten f_4, truly a "mortal sin").

4.1.2 The Kronecker Symbol or the "Discrete δ-Function"

Before we get into the definition of the discrete Fourier transformation (forward and inverse transformation), a few preliminary remarks are in order. From the continuous Fourier transformation $e^{i\omega t}$ we get for discrete times $t_k = k\,\Delta t$, $k = 0, 1, \ldots, N-1$ with $T = N\,\Delta t$:

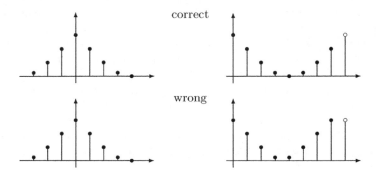

Fig. 4.1 Correctly wrapped-around (*top*); incorrectly wrapped-around (*bottom*)

$$\mathrm{e}^{\mathrm{i}\omega t} \rightarrow \mathrm{e}^{\mathrm{i}\frac{2\pi t_k}{T}} = \mathrm{e}^{\frac{2\pi \mathrm{i} k \Delta t}{N \Delta t}} = \mathrm{e}^{\frac{2\pi \mathrm{i} k}{N}} \equiv W_N^k. \tag{4.4}$$

Here the "kernel" is:

$$W_N = \mathrm{e}^{\frac{2\pi \mathrm{i}}{N}} \tag{4.5}$$

a very useful abbreviation. Occasionally we'll also need the discrete frequencies:

$$\omega_j = 2\pi j/(N \Delta t), \tag{4.6}$$

related to the discrete Fourier coefficients F_j (see below). The kernel W_N has the following properties:

$$W_N^{nN} = \mathrm{e}^{2\pi \mathrm{i} n} = 1 \qquad \text{for all integer } n,$$

$$W_N \text{ is periodic in } j \text{ and } k \text{ with the period } N. \tag{4.7}$$

A very useful representation of W_N may be obtained in the complex plane as a "clock-hand" in the unit circle.

The projection of the "hand of a clock" onto the real axis results in $\cos(2\pi n/N)$. Like when talking about a clock-face, we may, for example, call W_8^0 "3:00 a.m." or W_8^4 "9:00 a.m.". Now we can define the discrete "δ-function":

$$\sum_{j=0}^{N-1} W_N^{(k-k')j} = N \delta_{k,k'}. \tag{4.8}$$

Here $\delta_{k,k'}$ is the Kronecker symbol with the following property:

$$\delta_{k,k'} = \begin{cases} 1 \text{ for } k = k' \\ 0 \text{ else} \end{cases}. \tag{4.9}$$

This symbol (with prefactor N) accomplishes the same tasks the δ-function had when doing the continuous Fourier transformation. Equation (4.9) just means that, if the hand goes completely round the clock, we'll get zero, as we can see immediately by simply adding the hands' vectors in Fig. 4.2, except if the hand stops at "3:00 a.m.", a situation $k = k'$ can force. In this case we get N, as shown in Fig. 4.3.

Fig. 4.2 Representation of W_8^k in the complex plane

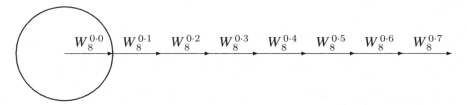

Fig. 4.3 For $N \to \infty$ (fictitious only) we quite clearly see the analogy with the δ-function

4.1.3 Definition of the Discrete Fourier Transformation

Now we want to determine the spectral content $\{F_j\}$ of the series $\{f_k\}$ using discrete Fourier transformation. For this purpose, we have to make the transition in the definition of the Fourier series:

$$c_j = \frac{1}{T} \int\limits_{-T/2}^{+T/2} f(t)e^{-2\pi ij/T}\,dt \tag{4.10}$$

with $f(t)$ periodic in T:

$$c_j = \frac{1}{N} \sum_{k=0}^{N-1} f_k e^{-2\pi ijk/N}. \tag{4.11}$$

In the exponent we find $\frac{k\Delta t}{N\Delta t}$, meaning that Δt can be eliminated. The prefactor contains the sampling raster Δt, so the prefactor becomes $\Delta t/T = \Delta t/(N\Delta t) = 1/N$. During the transition from (4.10) to (4.11) we tacitly shifted the limits of the interval from $-T/2$ to $+T/2$ to 0 to T, something that was okay, as we integrate over an *integer* period and $f(t)$ was assumed to be periodic in T. The sum has to come to an end at $N-1$, as this sampling point plus Δt reaches the limit of the interval. Therefore we get, for the discrete Fourier transformation:

Definition 4.1 *(Discrete Fourier transformation)*

$$F_j = \frac{1}{N} \sum_{k=0}^{N-1} f_k W_N^{-kj} \quad \text{with} \quad W_N = e^{2\pi i/N}. \tag{4.12}$$

The discrete *inverse* Fourier transformation is:

Definition 4.2 *(Discrete inverse Fourier transformation)*

$$f_k = \sum_{j=0}^{N-1} F_j W_N^{+kj} \quad \text{with} \quad W_N = e^{2\pi i/N}. \tag{4.13}$$

Please note that the inverse Fourier transformation doesn't have a prefactor $1/N$.

A bit of a warning is called for here. Instead of (4.12) and (4.13) we also come across definition equations with *positive* exponents for the *forward transformation* and with *negative* exponent for the *inverse transformation* (for example in "Numerical Recipes" [7]). This doesn't matter as far as the real part of $\{F_j\}$ is concerned. The imaginary part of $\{F_j\}$, however, changes its sign. Because we want to be consistent with the previous definitions of Fourier series and the continuous Fourier transformation we'd rather stick with the definitions (4.12) and (4.13) and remember that, for example, a *negative*, purely imaginary Fourier coefficient F_j belongs to a positive amplitude of a sine wave (given positive frequencies), as i of the forward transformation multiplied by i of the inverse transformation results in precisely a change of sign $i^2 = -1$. Often also the prefactor $1/N$ of the forward transformation is missing (for example in "Numerical Recipes" [7]). In view of the fact that F_0 is to be equal to the average of all samples, the prefactor $1/N$ really has to stay there, too. As we'll see, also "Parseval's theorem" will be grateful if we took care with our definition of the forward transformation. Using relation (4.8) we can see straight away that the inverse transformation (4.13) is correct:

$$f_k = \sum_{j=0}^{N-1} F_j W_N^{+kj} = \sum_{j=0}^{N-1} \frac{1}{N} \sum_{k'=0}^{N-1} f_{k'} W_N^{-k'j} W_N^{+kj}$$

$$= \frac{1}{N} \sum_{k'=0}^{N-1} f_{k'} \sum_{j=0}^{N-1} W_N^{(k-k')j} = \frac{1}{N} \sum_{k'=0}^{N-1} f_{k'} N \delta_{k,k'} = f_k. \tag{4.14}$$

Before we get into more rules and theorems, let's look at a few examples to illustrate the discrete Fourier transformation (Fig. 4.4)!

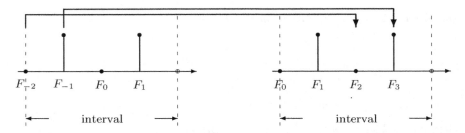

Fig. 4.4 Fourier coefficients with negative indices are wrapped to the right end of the interval

Example 4.1 ("Constant" with N = 4)

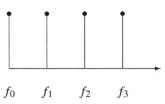

$$f_k = 1 \quad \text{for } k = 0, 1, 2, 3.$$

For the continuous Fourier transformation we expect a δ-function with the frequency $\omega = 0$. The discrete Fourier transformation therefore will only result in $F_0 \neq 0$. Indeed, we do get, using (4.12)—or even a lot smarter using (4.8):

$$\begin{aligned} F_0 &= \tfrac{1}{4}4 = 1 \\ F_1 &= 0 \\ F_2 &= 0 \\ F_3 &= 0. \end{aligned}$$

As $\{f_k\}$ is an even series, $\{F_j\}$ contains no imaginary part. The inverse transformation results in:

$$f_k = 1 \cos\left(2\pi \frac{k}{4}\underset{\underset{j=0}{\uparrow}}{0}\right) = 1 \quad \text{for } k = 0, 1, 2, 3.$$

Example 4.2 ("Cosine" with N = 4)

$$\begin{aligned} f_0 &= 1 \\ f_1 &= 0 \\ f_2 &= -1 \\ f_3 &= 0. \end{aligned}$$

We get, using (4.12) and $W_4 = i$:

$F_0 = 0$ (average = 0!)

$$F_1 = \frac{1}{4}(1 + (-1)(\text{"9:00 a.m."})) = \frac{1}{4}(1 + (-1)(-1)) = \frac{1}{2}$$

$$F_2 = \frac{1}{4}(1 + (-1)(\text{"3:00 p.m."})) = \frac{1}{4}(1 + (-1)1) \quad = 0$$

$$F_3 = \frac{1}{4}(1 + (-1)(\text{"9:00 p.m."})) = \frac{1}{4}(1 + (-1)(-1)) = \frac{1}{2}.$$

I bet you would have noticed that, due to the *negative* sign in the exponent in (4.12), we're running around *"clockwise"*. Maybe those of you who'd rather use a *positive* sign here, are *"Bavarians"*, who are well known for their clocks going backwards (you can actually buy them in souvenir-shops). So whoever uses a *plus* sign in (4.12) is out of sync with the rest of the world! What's $F_3 = 1/2$? Is there another spectral component, apart from the fundamental frequency $\omega_1 = 2\pi \times 1/4 \times \Delta t = \pi/(2\Delta t)$? Yes, there is! Of course it's the component with $-\omega_1$, that has been wrapped-around.

We can see that the *negative* frequencies of F_{N-1} (corresponding to smallest, not disappearing frequency ω_{-1}) are located from the right end of the interval decreasing to the left till they reach the center of the interval.

For *real* input the following applies:

$$F_{N-j} = F_j^*, \tag{4.15}$$

as we can easily deduce from (4.12). So in the case of *even* input the right half has exactly the same content as the left half; in the case of *odd* input, the right half will contain the conjugate complex or the same times minus as the left half. If we add together the intensity F_1 and $F_3 = F_{-1}$ shared "between brothers", this results in 1, as required by the input:

$$f_k = \frac{1}{2}i^k + \frac{1}{2}i^{3k} = \cos\left(2\pi\frac{k}{4}\right) \quad \text{for } k = 0, 1, 2, 3.$$

Example 4.3 ("Sine" with N = 4)

$$\begin{aligned} f_0 &= \quad 0 \\ f_1 &= \quad 1 \\ f_2 &= \quad 0 \\ f_3 &= -1. \end{aligned}$$

Again we use (4.12) and get:

$F_0 = 0$ (average $= 0$)

$F_1 = \frac{1}{4}(1 \times$ "6.00 a.m." $+(-1) \times$ "12.00 noon") $= \frac{1}{4}(-\mathrm{i}$ $+(-1) \times \mathrm{i}) = -\frac{\mathrm{i}}{2}$

$F_2 = \frac{1}{4}(1 \times$ "9.00 a.m." $+(-1) \times$ "9.00 p.m.") $= \frac{1}{4}(1 \times (-1) +(-1)(-1)) =$ 0

$F_3 = \frac{1}{4}(1 \times$ "12.00 noon" $+(-1) \times$ $\underbrace{\text{"6.00 a.m."}}_{\text{following day}}) = \frac{1}{4}(1 \times \mathrm{i}$ $+(-1)(-\mathrm{i})) =$ $\frac{\mathrm{i}}{2}$

real part $= 0$ imaginary part:

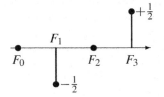

If we add the intensity with a minus sign for negative frequencies, that resulted from the sharing "between sisters", to the one for positive frequencies, meaning $F_1 + (-1)F_3 = -\mathrm{i}$, we get for the intensity of the sine wave (the inverse transformation provides us with another i!) the value 1:

$$f_k = -\frac{\mathrm{i}}{2}\mathrm{i}^k + \frac{\mathrm{i}}{2}\mathrm{i}^{3k} = \sin\left(2\pi\frac{k}{4}\right).$$

4.2 Theorems and Rules

4.2.1 Linearity Theorem

If we combine in a linear way $\{f_k\}$ and its series $\{F_j\}$ with $\{g_k\}$ and its series $\{G_j\}$, the we get:

$$\{f_k\} \leftrightarrow \{F_j\},$$
$$\{g_k\} \leftrightarrow \{G_j\}, \qquad\qquad (4.16)$$
$$a \cdot \{f_k\} + b \cdot \{g_k\} \leftrightarrow a \cdot \{F_j\} + b \cdot \{G_j\}.$$

Please always keep in mind that the discrete Fourier transformation contains only linear operators (in fact, basic maths only), but that the power representation is *no* linear operation.

4.2.2 The First Shifting Rule (Shifting in the Time Domain)

$$\{f_k\} \leftrightarrow \{F_j\}$$
$$\{f_{k-n}\} \leftrightarrow \{F_j W_N^{-jn}\}, \qquad n \text{ integer.} \tag{4.17}$$

A shift in the time domain by n results in a multiplication by the phase factor W_N^{-jn}.

Proof (First Shifting Rule)

$$
\begin{aligned}
F_j^{\text{shifted}} &= \frac{1}{N} \sum_{k=0}^{N-1} f_{k-n} W_N^{-kj} \\
&= \frac{1}{N} \sum_{k'=-n}^{N-1-n} f_{k'} W_N^{-(k'+n)j} \qquad \text{with } k - n = k' \tag{4.18} \\
&= \frac{1}{N} \sum_{k'=0}^{N-1} f_{k'} W_N^{-k'j} W_N^{-nj} = F_j^{\text{old}} W_N^{-nj}. \quad \square
\end{aligned}
$$

Because of the periodicity of f_k, we may shift the lower and the upper summation boundaries by n without a problem.

Example 4.4   (Shifted cosine with $N = 2$)

$$\{f_k\} = \{0, \ 1\} \qquad \text{or}$$
$$f_k = \frac{1}{2}\left(1 - \cos 2\pi k\right), \qquad k = 0, 1$$
$$W_2 = e^{i\pi} = -1$$
$$F_0 = \frac{1}{2}(0 + 1) = \frac{1}{2} \quad \text{(average)}$$
$$F_1 = \frac{1}{2}(0 + 1(-1)) = -\frac{1}{2} \qquad \text{consequently}$$
$$\{F_j\} = \left\{\frac{1}{2}, \ -\frac{1}{2}\right\}.$$

Now we shift the input by $n = 1$:

$$\{f_k^{\text{shifted}}\} = \{1, \ 0\} \qquad \text{or}$$
$$f_k = \frac{1}{2}\left(1 + \cos 2\pi k\right), \qquad k = 0, 1$$
$$\{F_j^{\text{shifted}}\} = \left\{\frac{1}{2}W_2^{-1\times 0}, \ \frac{1}{2}W_2^{-1\times 1}\right\} = \left\{\frac{1}{2}, \ \frac{1}{2}\right\}.$$

4.2.3 The Second Shifting Rule (Shifting in the Frequency Domain)

$$\{f_k\} \leftrightarrow \{F_j\}$$
$$\{f_k W_N^{-nk}\} \leftrightarrow \{F_{j+n}\}, \qquad n \text{ integer.} \tag{4.19}$$

A modulation in the time domain with W_N^{-nk} corresponds to a shift in the frequency domain. The proof is trivial.

Example 4.5 (Modulated cosine with $N = 2$)

$$\{f_k\} = \{0, \ 1\} \qquad \text{or}$$
$$f_k = \frac{1}{2}\left(1 - \cos \pi k\right), \qquad k = 0,1$$
$$\{F_j\} = \left\{\frac{1}{2}, \ -\frac{1}{2}\right\}.$$

Now we modulate the input with W_N^{-nk} with $n = 1$, that's $W_2^{-k} = (-1)^{-k}$, and get:

$$\{f_k^{\text{shifted}}\} = \{0, \ -1\} \qquad \text{or}$$
$$f_k = \frac{1}{2}\left(-1 + \cos \pi k\right), \qquad k = 0,1$$
$$\{F_j^{\text{shifted}}\} = \{F_{j-1}\} = \left\{-\frac{1}{2}, \ \frac{1}{2}\right\}.$$

Here, F_{-1} was wrapped to $F_{2-1} = F_1$.

4.2.4 Scaling Rule/Nyquist Frequency

From Fig. 4.5 we see that the highest frequency ω_{max} or also $-\omega_{\text{max}}$ corresponds to the center of the series of Fourier coefficients. This we get by inserting $j = N/2$ in (4.6):

$$\Omega_{\text{Nyq}} = \frac{\pi}{\Delta t} \qquad \text{"Nyquist frequency".} \tag{4.20}$$

This frequency often is also called the cut-off frequency. If we take a sample, say every μs ($\Delta t = 10^{-6}$ s), then Ω_{Nyq} is 3.14 megaradians/s (if you prefer to think in frequencies instead of angular frequencies: $\nu_{\text{Nyq}} = \Omega_{\text{Nyq}}/2\pi$, so here 0.5 MHz).

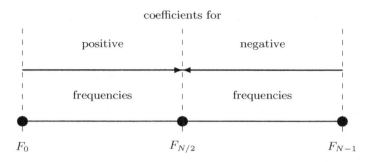

Fig. 4.5 Positioning of the Fourier coefficients

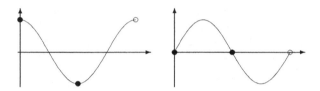

Fig. 4.6 Two samples per period: cosine (*left*); sine (*right*)

So the Nyquist frequency Ω_{Nyq} corresponds to taking *two* samples per period, as shown in Fig. 4.6.

While we'll get away with this in the case of the cosine, by the skin of our teeth, it definitely won't work for the sine! Here we grabbed the samples at the wrong moment, or maybe there was no signal after all (for example because a cable hadn't been plugged in, or due to a power cut). In fact, the imaginary part of f_k at the Nyquist frequency always is 0. The Nyquist frequency therefore is the highest possible spectral component for a cosine wave; for the sine it's only up to:

$$\omega = 2\pi(N/2 - 1)/(N\Delta t) = \Omega_{\text{Nyq}}(1 - 2/N).$$

Equation (4.20) is our scaling theorem, as the choice of Δt allows us to stretch or compress the time axis, while keeping the number of samples N constant. This only has an impact on the frequency scale running from $\omega = 0$ to $\omega = \Omega_{\text{Nyq}}$. Δt doesn't appear anywhere else!

The normalisation factor we came across in (1.41) and (2.32), is done away with here, as using discrete Fourier transformation we normalise to the number of samples N, regardless of the sampling raster Δt.

4.3 Convolution, Cross Correlation, Autocorrelation, Parseval's Theorem

Before we're able to formulate the discrete versions of the (2.34), (2.48), (2.52), and (2.54), we have to get a handle on two problems:

i. The number of samples N for the two functions $f(t)$ and $g(t)$ we want to convolve or cross-correlate, must be the same. This often is not the case, for example if $f(t)$ is the "theoretical" signal we would get for a δ-shaped instrumental resolution function, which, however, has to be convolved with the finite resolution function $g(t)$. There's a simple fix: we pad the series $\{g_k\}$ with zeros so we get N samples, just like in the case of series $\{f_k\}$.

ii. Don't forget, that $\{f_k\}$ is periodic in N and our "padded" $\{g_k\}$, too. This means that negative indices are wrapped-around to the right end of the interval. The resolution function $g(t)$ mentioned in Fig. 4.7, which we assumed to be symmetrical, had 3 samples and got padded with 5 zeros to a total of $N = 8$ and is displayed in Fig. 4.7.

Another extreme example:

Example 4.6 (Rectangle) We'll remember that a continuous "rectangular function", when convolved with itself in the interval $-T/2 \le t \le +T/2$, results in a "triangular function" in the interval $-T \le t \le +T$. In the discrete case, the "triangle" gets wrapped in the area $-T \le t \le -T/2$ to $0 \le t \le T/2$. The same happens to the "triangle" in the area $+T/2 \le t \le +T$, which gets wrapped to $-T/2 \le t \le 0$. Therefore both halves of the interval are "corrupted" by the wrap-around, so that the end-result is another constant (cf. Fig. 4.8). No wonder! This "rectangular function" with *periodic* continuation is a constant! And a constant convolved with a constant naturally is another constant.

As long as $\{f_k\}$ is periodic in N, there's nothing wrong with the fact that upon convolution data from the end/beginning of the interval will be "mixed into" data from the beginning/end of the interval. If you don't like that—for whatever reasons— rather also pad $\{f_k\}$ with zeros, using precisely the correct number of zeros so $\{g_k\}$ won't create overlap between f_0 and f_{N-1} any more.

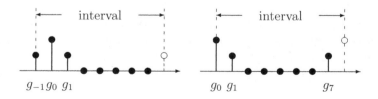

Fig. 4.7 Resolution function $\{g_k\}$: without wrap-around (*left*); with wrap-around (*right*)

Fig. 4.8 Convolution of a "rectangular function" with itself: without wrap-around (*top*); with wrap-around (*bottom*)

4.3.1 Convolution

We'll define the discrete convolution as follows:

Definition 4.3 (*Discrete convolution*)

$$h_k \equiv (f \otimes g)_k = \frac{1}{N} \sum_{l=0}^{N-1} f_l g_{k-l}. \tag{4.21}$$

The "convolution sum" is commutative, distributive and associative. The normalisation factor $1/N$ in context: the convolution of $\{f_k\}$ with the "discrete δ-function" $\{g_k\} = N\delta_{k,0}$ is to leave the series $\{f_k\}$ unchanged. Following this rule, also a "normalised" resolution function $\{g_k\}$ should respect the condition $\sum_{k=0}^{N-1} g_k = N$. Unfortunately often the convolution also gets defined without the prefactor $1/N$.

The Fourier transform of $\{h_k\}$ is:

$$
\begin{aligned}
H_j &= \frac{1}{N} \sum_{k=0}^{N-1} \frac{1}{N} \sum_{l=0}^{N-1} f_l g_{k-l} W_N^{-kj} \\
&= \frac{1}{N^2} \sum_{k=0}^{N-1} \sum_{l=0}^{N-1} f_l W_N^{-lj} g_{k-l} W_N^{-kj} W_N^{+lj} \\
&\qquad\qquad \uparrow \quad \text{extended} \quad \uparrow \\
&= \frac{1}{N^2} \sum_{l=0}^{N-1} f_l W_N^{-lj} \sum_{k'=-l}^{N-1-l} g_{k'} W_N^{-k'j} \qquad \text{with } k' = k - l \\
&= F_j G_j.
\end{aligned}
\tag{4.22}
$$

In our last step we took advantage of the fact that, due to the periodicity in N, the second sum may also run from 0 to $N - 1$ instead of $-l$ to $N - 1 - l$. This, however, makes sure that the current index l has been totally eliminated from the second sum, and we get the product of the Fourier transform F_j and G_j. So we arrive at the discrete Convolution Theorem:

$$
\begin{aligned}
\{f_k\} &\leftrightarrow \{F_j\}, \\
\{g_k\} &\leftrightarrow \{G_j\}, \\
\{h_k\} = \{(f \otimes g)_k\} &\leftrightarrow \{H_j\} = \{F_j \cdot G_j\}.
\end{aligned}
\tag{4.23}
$$

The convolution of the series $\{f_k\}$ and $\{g_k\}$ results in a product in the Fourier space.

The inverse Convolution Theorem is:

$$
\begin{aligned}
\{f_k\} &\leftrightarrow \{F_j\}, \\
\{g_k\} &\leftrightarrow \{G_j\}, \\
\{h_k\} = \{f_k \cdot g_k\} &\leftrightarrow \{H_j\} = \{N(F \otimes G)_j\}.
\end{aligned}
\tag{4.24}
$$

Proof (Inverse Convolution Theorem)

$$
H_j = \frac{1}{N} \sum_{k=0}^{N-1} f_k g_k W_N^{-kj} = \frac{1}{N} \sum_{k=0}^{N-1} f_k g_k \underbrace{\sum_{k'=0}^{N-1} W_N^{-k'j} \delta_{k,k'}}_{k'\text{-sum ``artificially'' introduced}}
$$

$$
= \frac{1}{N^2} \sum_{k=0}^{N-1} f_k \sum_{k'=0}^{N-1} g_{k'} W_N^{-k'j} \underbrace{\sum_{l=0}^{N-1} W_N^{-l(k-k')}}_{l\text{-sum yields } N\delta_{k,k'}}
$$

$$
= \sum_{l=0}^{N-1} \frac{1}{N} \sum_{k=0}^{N-1} f_k W_N^{-lk} \frac{1}{N} \sum_{k'=0}^{N-1} g_{k'} W_N^{-k'(j-l)}
$$

$$
= \sum_{l=0}^{N-1} F_l G_{j-l} = N(F \otimes G)_j. \quad \square
$$

Example 4.7 (Nyquist frequency with N = 8)

$\{f_k\} = \{1, \ 0, \ 1, \ 0, \ 1, \ 0, \ 1, \ 0\},$

$\{g_k\} = \{4, \ 2, \ 0, \ 0, \ 0, \ 0, \ 0, \ 2\}.$

The "resolution function" $\{g_k\}$ is padded to $N = 8$ with zeros and normalised to $\sum_{k=0}^{7} g_k = 8$. The convolution of $\{f_k\}$ with $\{g_k\}$ results in:

$$\{h_k\} = \left\{\frac{1}{2}, \ \frac{1}{2}, \ \frac{1}{2}, \ \frac{1}{2}, \ \frac{1}{2}, \ \frac{1}{2}, \ \frac{1}{2}, \ \frac{1}{2}\right\},$$

meaning, that everything gets "flattened", because the resolution function (here triangle-shaped) has a full half-width of Δt and consequently doesn't allow the recording of oscillations with the period Δt. The Fourier transform therefore is $H_k = 1/2\delta_{k,0}$. Using the Convolution Theorem (4.23) we would get:

$$\{F_j\} = \left\{\frac{1}{2}, \ 0, \ 0, \ 0, \ \frac{1}{2}, \ 0, \ 0, \ 0\right\}.$$

The result is easy to understand: the average is $1/2$, at the Nyquist frequency we have $1/2$, all other elements are 0.

$$G_0 = 1 \qquad \left(\frac{1}{8} \times \text{average}\right)$$

$$G_1 = \frac{1}{2} + \frac{\sqrt{2}}{4} \quad \left(\frac{1}{8}\{4 + 2 \times \text{``4:30 a.m.''} + 2 \times \text{``1:30 p.m.''}\}\right)$$

$$G_2 = \frac{1}{2} \qquad \left(\frac{1}{8}\{4 + 2 \times \text{``6:00 a.m.''} + 2 \times \text{``12:00 midnight''}\}\right)$$

$$G_3 = \frac{1}{2} - \frac{\sqrt{2}}{4} \quad \left(\frac{1}{8}\{4 + 2 \times \text{``7:30 a.m.''} + 2 \times \text{``10:30 a.m. next day''}\}\right)$$

$$G_4 = 0 \qquad \left(\frac{1}{8}\{4 + 2 \times \text{``9:00 a.m.''} + 2 \times \text{``9:00 p.m. next day''}\}\right)$$

$$G_5 = \frac{1}{2} - \frac{\sqrt{2}}{4}$$

$$G_6 = \frac{1}{2} \qquad \Bigg\} \quad \text{because of real input,}$$

$$G_7 = \frac{1}{2} + \frac{\sqrt{2}}{4}$$

hence:

$$\{G_j\} = \left\{ 1, \ \frac{1}{2} + \frac{\sqrt{2}}{4}, \ \frac{1}{2}, \ \frac{1}{2} - \frac{\sqrt{2}}{4}, \ 0, \ \frac{1}{2} - \frac{\sqrt{2}}{4}, \ \frac{1}{2}, \ \frac{1}{2} + \frac{\sqrt{2}}{4} \right\}.$$

For the product we get $H_j = F_j G_j = \{1/2, 0, 0, 0, 0, 0, 0, 0\}$, like we should for the Fourier transform. If we'd taken the Convolution Theorem seriously right from the beginning, then the calculation of G_0 (average) and G_4 at the Nyquist frequency would have been quite sufficient, as all other $F_j = 0$. The fact that the Fourier transform of the resolution function for the Nyquist frequency is 0, precisely means that with this resolution function we're not able to record oscillations with the Nyquist frequency any more. Our inputs, however, were only the frequency 0 and the Nyquist frequency.

4.3.2 Cross Correlation

We define for the discrete cross correlation between $\{f_k\}$ and $\{g_k\}$, similar to what we did in (2.48):

Definition 4.4 *(Discrete cross correlation)*

$$h_k \equiv (f \star g)_k = \frac{1}{N} \sum_{l=0}^{N-1} f_l \cdot g_{l+k}^*. \tag{4.25}$$

If the indices at g_k go beyond $N-1$, then we'll simply subtract N (periodicity). The cross correlation between $\{f_k\}$ and $\{g_k\}$, of course, results in a product of their Fourier transforms:

$$\begin{aligned}
\{f_k\} &\leftrightarrow \{F_j\}, \\
\{g_k\} &\leftrightarrow \{G_j\}, \\
\{h_k\} = \{(f \star g)_k\} &\leftrightarrow \{H_j\} = \{F_j \cdot G_j^*\}.
\end{aligned} \tag{4.26}$$

Proof (Discrete cross correlation)

$$H_j = \frac{1}{N} \sum_{k=0}^{N-1} \frac{1}{N} \sum_{l=0}^{N-1} f_l g_{l+k}^* W_N^{-kj}$$

$$= \frac{1}{N} \sum_{l=0}^{N-1} f_l \frac{1}{N} \sum_{k=0}^{N-1} g_{l+k}^* W_N^{-kj}$$

with the First Shifting Rule and complex conjugate

$$= \frac{1}{N} \sum_{l=0}^{N-1} f_l G_j^* W_N^{-jl} = F_j G_j^*. \quad \square$$

4.3.3 Autocorrelation

Here we have $\{f_k\} = \{g_k\}$, which leads to:

$$h_k \equiv (f \star f)_k = \frac{1}{N} \sum_{l=0}^{N-1} f_l \cdot f_{l+k}^* \tag{4.27}$$

and:

$$\begin{aligned} \{f_k\} &\leftrightarrow \{F_j\}, \\ \{h_k\} = \{(f \star f)_k\} &\leftrightarrow \{H_j\} = \{|F_j|^2\}. \end{aligned} \tag{4.28}$$

In other words: the Fourier transform of the autocorrelation of $\{f_k\}$ is the modulus squared of the Fourier series $\{F_j\}$ or its power representation.

4.3.4 Parseval's Theorem

We use (4.27) for $k = 0$, that's h_0 ("without time-lag"), and get on the one side:

$$h_0 = \frac{1}{N} \sum_{l=0}^{N-1} |f_l|^2. \tag{4.29}$$

On the other side, the inverse transformation of $\{H_j\}$, especially for $k = 0$, results in (cf. 4.13):

$$h_0 = \sum_{j=0}^{N-1} |F_j|^2. \tag{4.30}$$

Put together, this gives us the discrete version of Parseval's theorem:

$$\frac{1}{N} \sum_{l=0}^{N-1} |f_l|^2 = \sum_{j=0}^{N-1} |F_j|^2. \tag{4.31}$$

Example 4.8 ("Parseval's theorem" for $N = 2$)

$\{f_l\} = \{0, \ 1\}$ (cf. example for First Shifting Rule
Sect. 4.2.2)

$\{F_j\} = \{1/2, \ -1/2\}$ (here there is only the average F_0
and the Nyquist frequency at F_1!)

$$\frac{1}{2} \sum_{l=0}^{N} |f_l|^2 = \frac{1}{2} \times 1 = \frac{1}{2}$$

$$\sum_{j=0}^{N} |F_j|^2 = \frac{1}{4} + \frac{1}{4} = \frac{1}{2}.$$

Caution: Often the prefactor $1/N$ gets left out when defining Parseval's theorem. To stay consistent with all other definitions, however, it should not be missing here!

4.4 The Sampling Theorem

When discussing the Nyquist frequency, we already mentioned that we need at least two samples per period to show cosine oscillations at the Nyquist frequency. Now we'll turn the tables and claim that as a matter of principle we won't be looking at anything but functions $f(t)$ that are "bandwidth-limited", meaning, that outside the interval $[-\Omega_{\mathrm{Nyq}}, \ \Omega_{\mathrm{Nyq}}]$ their Fourier transforms $F(\omega)$ are 0. In other words: we'll refine our sampling to a degree where we just manage to capture all the spectral components of $f(t)$. This seemingly "innocent" requirement implies that the periodically continued function and all its derivatives are continuous everywhere, such that its Fourier series is finite. Now we'll skilfully "marry" formulas we've learned when dealing with the Fourier series expansion and the continuous Fourier transformation with each other, and then pull the sampling theorem out of the hat. For this purpose we'll recall (1.26) and (1.27) which show that a periodic function $f(t)$ can be expanded into an (infinite) Fourier series:

$$f(t) = \sum_{k=-\infty}^{+\infty} C_k e^{i2\pi kt/T}$$

$$\text{with } C_k = \frac{1}{T} \int_{-T/2}^{T/2} f(t) e^{-i2\pi kt/T} dt.$$

Since $F(\omega)$ is zero outside $[-\Omega_{\text{Nyq}}, \Omega_{\text{Nyq}}]$ we can continue this function periodically and expand it into an infinite Fourier series. So we replace: $f(t) \rightarrow F(\omega)$, $t \rightarrow \omega$, $T/2 \rightarrow \Omega_{\text{Nyq}}$ and get:

$$F(\omega) = \sum_{k=-\infty}^{+\infty} C_k e^{i\pi k\omega/\Omega_{\text{Nyq}}}$$

$$\text{with } C_k = \frac{1}{2\Omega_{\text{Nyq}}} \int_{-\Omega_{\text{Nyq}}}^{+\Omega_{\text{Nyq}}} F(\omega) e^{-i\pi k\omega/\Omega_{\text{Nyq}}} d\omega. \tag{4.32}$$

A similar integral also occurs in the defining equation for the inverse continuous Fourier transformation:

$$f(t) = \frac{1}{2\pi} \int_{-\Omega_{\text{Nyq}}}^{+\Omega_{\text{Nyq}}} F(\omega) e^{i\omega t} d\omega. \tag{4.33}$$

The integrations boundaries are $\pm\Omega_{\text{Nyq}}$, as $F(\omega)$ is bandwidth-limited. When we compare this with (4.32) we get:

$$2\Omega_{\text{Nyq}} C_k = 2\pi f(-\pi k/\Omega_{\text{Nyq}}). \tag{4.34}$$

Once we've inserted this in (4.32) we get:

$$F(\omega) = \frac{\pi}{\Omega_{\text{Nyq}}} \sum_{k=-\infty}^{+\infty} f(-\pi k/\Omega_{\text{Nyq}}) e^{i\pi k\omega/\Omega_{\text{Nyq}}}. \tag{4.35}$$

When we finally insert this into the defining equation (4.33), we get:

$$f(t) = \frac{1}{2\pi} \int_{-\Omega_{\text{Nyq}}}^{+\Omega_{\text{Nyq}}} \frac{\pi}{\Omega_{\text{Nyq}}} \sum_{k=-\infty}^{+\infty} f\left(\frac{-\pi k}{\Omega_{\text{Nyq}}}\right) e^{i\pi k\omega/\Omega_{\text{Nyq}}} e^{i\omega t} d\omega$$

$$= \frac{1}{2\Omega_{\text{Nyq}}} \sum_{k=-\infty}^{+\infty} f(-k\Delta t) 2 \int_0^{+\Omega_{\text{Nyq}}} \cos \omega(t + k\Delta t) d\omega \qquad (4.36)$$

$$= \frac{1}{2\Omega_{\text{Nyq}}} \sum_{k=-\infty}^{+\infty} f(-k\Delta t) 2 \frac{\sin \Omega_{\text{Nyq}}(t + k\Delta t)}{(t + k\Delta t)}.$$

By replacing $k \to -k$ (it's not important in which order the sums are calculated) we get the Sampling Theorem:

$$\text{Sampling Theorem: } f(t) = \sum_{k=-\infty}^{+\infty} f(k\Delta t) \frac{\sin \Omega_{\text{Nyq}}(t - k\Delta t)}{\Omega_{\text{Nyq}}(t - k\Delta t)}. \qquad (4.37)$$

In other words, we can reconstruct the function $f(t)$ for *all* times t from the samples at the times $k\Delta t$, provided the function $f(t)$ is "bandwidth-limited", i.e. it contains no frequencies above $\Omega_{\text{Nyq}} = \pi/\Delta t$. To achieve this, we only need to multiply $f(k\Delta t)$ with the function $\frac{\sin x}{x}$ (with $x = \Omega_{\text{Nyq}}(t - k\Delta t)$) and sum up over all samples. The factor $\frac{\sin x}{x}$ naturally is equal to 1 for $t = k\Delta t$, for other times, $\frac{\sin x}{x}$ decays and slowly oscillates towards zero, which means, that $f(t)$ is a composite of plenty of $\left(\frac{\sin x}{x}\right)$-functions at the location $t = k\Delta t$ with the amplitude $f(k\Delta t)$. Note that for adequate sampling with $\Delta t = \frac{\pi}{\Omega_{\text{Nyq}}}$ each k-term in the sum in (4.37) contributes $f(k\Delta t)$ at the sampling points $t = k\Delta t$ and zero at all other sampling points whereas all terms contribute to the interpolation between sampling points.

Example 4.9 (Sampling Theorem with $N = 2$)

$$f_0 = 1$$
$$f_1 = 0.$$

We expect:

$$f(t) = \frac{1}{2} + \frac{1}{2} \cos \Omega_{\text{Nyq}} t = \cos^2 \frac{\Omega_{\text{Nyq}} t}{2}.$$

The sampling theorem tells us:

$$f(t) = \sum_{k=-\infty}^{+\infty} f_k \frac{\sin \Omega_{\text{Nyq}}(t - k\Delta t)}{\Omega_{\text{Nyq}}(t - k\Delta t)}$$

with $f_k = \delta_{k,\text{even}}$ (with periodic continuation)

$$= \frac{\sin \Omega_{\text{Nyq}} t}{\Omega_{\text{Nyq}} t} + \sum_{l=1}^{+\infty} \frac{\sin \Omega_{\text{Nyq}}(t - 2l\Delta t)}{\Omega_{\text{Nyq}}(t - 2l\Delta t)} + \sum_{l=1}^{+\infty} \frac{\sin \Omega_{\text{Nyq}}(t + 2l\Delta t)}{\Omega_{\text{Nyq}}(t + 2l\Delta t)}$$

with the substitution $k = 2l$

$$= \frac{\sin \Omega_{\text{Nyq}} t}{\Omega_{\text{Nyq}} t} + \sum_{l=1}^{+\infty} \left[\frac{\sin 2\pi \left(\frac{t}{2\Delta t} - l \right)}{2\pi \left(\frac{t}{2\Delta t} - l \right)} + \frac{\sin 2\pi \left(\frac{t}{2\Delta t} + l \right)}{2\pi \left(\frac{t}{2\Delta t} + l \right)} \right]$$

with $\Omega_{\text{Nyq}} \Delta t = \pi$

$$= \frac{\sin \Omega_{\text{Nyq}} t}{\Omega_{\text{Nyq}} t} + \frac{1}{2\pi} \sum_{l=1}^{+\infty} \frac{\left(\frac{t}{2\Delta t} + l \right) \sin \Omega_{\text{Nyq}} t + \left(\frac{t}{2\Delta t} - l \right) \sin \Omega_{\text{Nyq}} t}{\left(\frac{t}{2\Delta t} - l \right) \left(\frac{t}{2\Delta t} + l \right)}$$

$$= \frac{\sin \Omega_{\text{Nyq}} t}{\Omega_{\text{Nyq}} t} + \frac{\sin \Omega_{\text{Nyq}} t}{2\pi} \frac{2t}{2\Delta t} \sum_{l=1}^{+\infty} \frac{1}{\left(\frac{t}{2\Delta t} \right)^2 - l^2}$$

$$= \frac{\sin \Omega_{\text{Nyq}} t}{\Omega_{\text{Nyq}} t} \left(1 + \left(\frac{\Omega_{\text{Nyq}} t}{2\pi} \right)^2 2 \sum_{l=1}^{+\infty} \frac{1}{\left(\frac{\Omega_{\text{Nyq}} t}{2\pi} \right)^2 - l^2} \right)$$

with [8, No 1.421.3]

$$= \frac{\sin \Omega_{\text{Nyq}} t}{\Omega_{\text{Nyq}} t} \pi \frac{\Omega_{\text{Nyq}} t}{2\pi} \cot \frac{\pi \Omega_{\text{Nyq}} t}{2\pi}$$

$$= \sin \Omega_{\text{Nyq}} t \frac{1}{2} \frac{\cos(\Omega_{\text{Nyq}} t / 2)}{\sin(\Omega_{\text{Nyq}} t / 2)}$$

$$= 2 \sin(\Omega_{\text{Nyq}} t / 2) \cos(\Omega_{\text{Nyq}} t / 2) \frac{1}{2} \frac{\cos(\Omega_{\text{Nyq}} t / 2)}{\sin(\Omega_{\text{Nyq}} t / 2)} = \cos^2 \left(\Omega_{\text{Nyq}} t / 2 \right). \quad (4.38)$$

Please note that we actually do need all summation terms of $k = -\infty$ to $k = +\infty$! If we had only taken $k = 0$ and $k = 1$ into consideration, we would have got:

$$f(t) = 1 \frac{\sin \Omega_{\text{Nyq}} t}{\Omega_{\text{Nyq}} t} + 0 \frac{\sin \Omega_{\text{Nyq}} (t - \Delta t)}{\Omega_{\text{Nyq}} (t - \Delta t)} = \frac{\sin \Omega_{\text{Nyq}} t}{\Omega_{\text{Nyq}} t}$$

which would not correspond to the input of $\cos^2(\Omega_{\text{Nyq}} t / 2)$. We still would have, as before, $f(0) = 1$ and $f(t = \Delta t) = 0$, but for $0 < t < \Delta t$, we wouldn't have interpolated correctly, as $\frac{\sin x}{x}$ slowly decays for big x, while we, however, want to get a periodic oscillation that doesn't decay as input. You will realise, that the sampling theorem—similar to Parseval's equation (1.50)—is good for the summation of certain infinite series.

What happens if, for some reason or other, our sampling happens to be too coarse and $F(\omega)$ above Ω_{Nyq} was unequal to 0? Quite simple: the spectral density above Ω_{Nyq} will be "reflected" to the interval $0 \leq \omega \leq \Omega_{\text{Nyq}}$, meaning that the true spectral density gets "corrupted" by the part that would be outside the interval.

Example 4.10 (Not enough samples) We'll take a cosine input and a bit less than two samples per period (cf. Fig. 4.9).

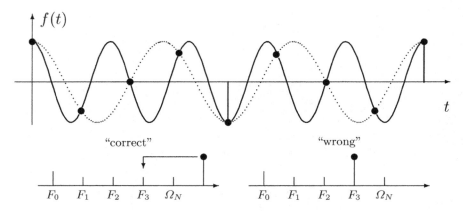

Fig. 4.9 Less than two samples per period (*top*): cosine input (*solid line*); "apparently" lower frequency (*dotted line*). Fourier coefficients with wrap-around (*bottom*)

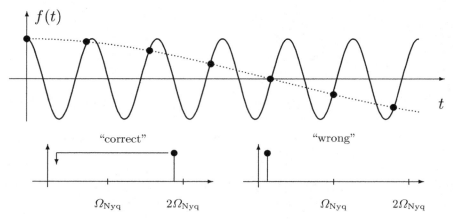

Fig. 4.10 Slightly more than one sample per period (*top*): cosine input (*solid line*); "apparently" lower frequency (*dotted line*). Fourier coefficients with wrap-around (*bottom*)

Here there are eight samples for five periods, and that means that Ω_{Nyq} has been exceeded by 25 %. The broken line in Fig. 4.9 shows that a function with only three periods would produce the same samples within the same interval.

Therefore the discrete Fourier transformation will show a lower spectral component, namely at $\Omega_{\mathrm{Nyq}} - 25$ %. This will become quite obvious, indeed, when we use only slightly more than one sample per period.

Here $\{F_j\}$ produces only a very low-frequency component (cf. Fig. 4.10). In other words: spectral density that would appear at $\approx 2\Omega_{\mathrm{Nyq}}$, appears at $\omega \approx 0$! This "corruption" of the spectral density through insufficient sampling is called "aliasing", similar to someone acting under an assumed name. In a nutshell: When sampling, rather err on the fine side than the coarse one! Coarser rasters can always be achieved later on by compressing data sets, but it will never work the other way round!

4.5 Data Mirroring

Often we have a situation where, on top of the samples $\{f_k\}$, we also know that the series starts with $f_0 = 0$ or at f_0 with horizontal tangent ($\stackrel{\wedge}{=}$ slope $= 0$). In this case we should use data mirroring forcing a situation where the input is an odd or an even series (cf. Fig. 4.11):

$$
\begin{aligned}
&\text{odd:}\\
&f_{2N-k} = -f_k \quad k = 0, 1, \ldots, N-1, \quad \text{here we put } f_N = 0;\\
&\text{even:}\\
&f_{2N-k} = +f_k \quad k = 0, 1, \ldots, N-1, \quad \text{here } f_N \text{ is undetermined!}
\end{aligned}
\tag{4.39}
$$

For odd series we put $f_N = 0$, as would be the case for periodic continuation anyway. For even series this is not necessarily the case. A possibility to determine f_N would be $f_N = f_0$ (as if we wanted to continue the non-mirrored data set periodically). In our example of Fig. 4.11 this would result in a δ-spike at f_N, which wouldn't make sense. Equally, in our example $f_N = 0$ can't be used (another δ-spike!). A better choice would be $f_N = f_{N-1}$, and even better $f_N = -f_0$ for the present case. The optimum choice, however, depends on the respective problem. So, for example, in the case of a cosine with window function and subsequently plenty of zeros, $f_N = 0$ would be the correct choice (cf. Fig. 4.12).

Now the interval is twice as long! Apply the normal fast Fourier transformation and you'll have a lot of fun with it, even if (or maybe exactly because of it?) the real part (in the case of odd mirroring) or the imaginary part (in the case of even

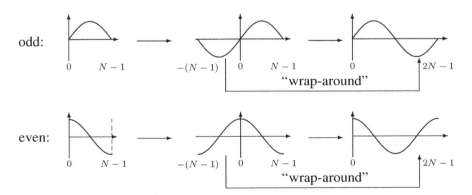

Fig. 4.11 Odd/even input, forced by data mirroring

Fig. 4.12 Example for the choice of f_N

Fig. 4.13 Basis functions for cosine- (*left*) and for sine-transformation (*right*)

mirroring) is full of zeros. If you don't like that, use a more efficient algorithm using the fast sine- or cosine-transformation.

As we can see in Fig. 4.13, for these sine- or cosine-transformations *other* basis functions are being used than the fundamental and harmonics of the normal Fourier transform, to model the input: also all functions with half the period will occur (the second half models the mirror image). The normal Fourier transformation of the mirrored input reads:

$$
F_j = \frac{1}{2N} \sum_{k=0}^{2N-1} f_k W_{2N}^{-kj} = \frac{1}{2N} \left(\sum_{k=0}^{N-1} f_k W_{2N}^{-kj} + \sum_{k=N}^{2N-1} f_k W_{2N}^{-kj} \right)
$$

$$
= \frac{1}{2N} \left(\sum_{k=0}^{N-1} f_k W_{2N}^{-kj} + \sum_{k'=N}^{1} f_{2N-k'} W_{2N}^{-(2N-k')j} \right)
$$

$$
\text{sequence irrelevant}
$$

$$
= \frac{1}{2N} \left(\sum_{k=0}^{N-1} f_k W_{2N}^{-kj} + \sum_{k'=1}^{N} (\pm) f_{k'} \underbrace{W_{2N}^{-2Nj}}_{} W_{2N}^{+k'j} \right)
$$

$$
\text{for} \begin{pmatrix} \text{even} \\ \text{odd} \end{pmatrix} \qquad = e^{-2\pi i \frac{2Nj}{2N}} = 1
$$

$$
= \frac{1}{2N} \left\{ \begin{pmatrix} 1 \\ -i \end{pmatrix} \sum_{k=0}^{N-1} f_k \times 2 \begin{pmatrix} \cos \frac{2\pi kj}{2N} \\ \sin \frac{2\pi kj}{2N} \end{pmatrix} + f_N e^{-i\pi j} - f_0 \right\}
$$

$$
= \begin{cases} \dfrac{1}{N} \displaystyle\sum_{k=0}^{N-1} f_k \cos \dfrac{\pi kj}{N} + \dfrac{1}{2N} \left(f_N e^{-i\pi j} - f_0 \right) & \text{even} \\[3ex] \dfrac{-i}{N} \displaystyle\sum_{k=0}^{N-1} f_k \sin \dfrac{\pi kj}{N} & \text{odd} \end{cases} \quad .
$$

The expressions $(1/N) \sum_{k=0}^{N-1} f_k \cos(\pi k j/N)$ and $(1/N) \sum_{k=0}^{N-1} f_k \sin(\pi k j/N)$ are called cosine- and sine-transformation. Please note:

i. The arguments for the cosine-/sine-function are $\pi k j/N$ and not $2\pi k j/N$! This shows, that half periods as basis function are also allowed (cf. Fig. 4.13).
ii. In the case of the sine transformation shifting of the sine boundaries from $k' = 1, 2, \ldots, N$ towards $k' = 0, 1, \ldots, N - 1$ is no problem, as the following has to be true: $f_N = f_0 = 0$. Apart from the factor $-i$ the sine transformation is identical to the normal Fourier transformation of the mirrored input, though it only has half as many coefficients. The inverse sine transformation is identical to the forward transformation, with the exception of the normalisation.
iii. In the case of the cosine transformation, the terms $(1/2N)(f_N e^{-i\pi j} - f_0)$ stay, except if they happen to be equal to 0. That means, that generally the cosine transformation will not be identical to the normal Fourier transformation of the mirrored input! (Fig. 4.14).
iv. Obviously Parseval's theorem does *not* apply to the cosine transformation.
v. Obviously the inverse cosine transformation is not identical to the forward transformation, apart from factors.

Example 4.11 ("Constant", $N = 4$)

$$\{f_k\} = 1 \qquad \text{for all } k.$$

The normal Fourier transformation of the mirrored input is:

$$F_0 = \frac{1}{8} 8 = 1, \qquad \text{all other } F_j = 0.$$

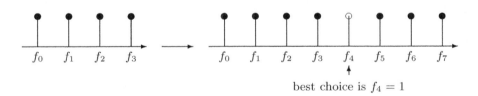

best choice is $f_4 = 1$

Fig. 4.14 Input without mirroring (*left*); with mirroring (*right*)

Cosine transformation:

$$F_j = \frac{1}{4} \sum_{k=0}^{3} \cos \frac{\pi k j}{4} = \begin{cases} \frac{1}{4}4 = 1 \text{ for } j = 0 \\ \frac{1}{4}\delta_{j,\text{odd}} \text{ for } j \neq 0 \end{cases}.$$

Here the flip-side is that, because of $\cos(\pi k j/N)$, the clock-hand or its projection onto the real axis only run around half as fast, and consequently relation (4.8) becomes false.

The extra terms can be omitted only if $f_0 = f_N = 0$ is true, as for example in Fig. 4.15.

If you insist on using the cosine transformation, "correct" it using the term:

$$\frac{1}{2N}(f_N e^{-i\pi j} - f_0).$$

Then you get the normal Fourier transformation of the mirrored data set, and no harm was done. In our above example, the one with the constant input, this would look as shown in Fig. 4.16.

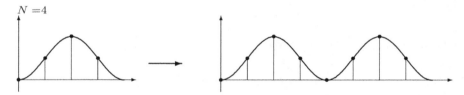

Fig. 4.15 Input (*left*); with correct mirroring (*right*)

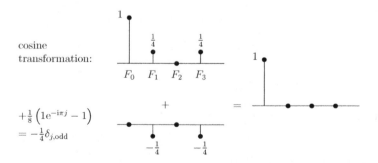

Fig. 4.16 Cosine transformation with correcting terms

4.6 How to Get Rid of the "Straight-Jacket" Periodic Continuation? By Using Zero-Padding!

So far, we had chosen all our examples in a way where $\{f_k\}$ could be continued periodically without a problem. For example, we truncated a cosine precisely where there was no problem continuing the cosine-shape periodically. In practice, this often can't be done:

i. we'd have to know the period in the first place to be able to know where to truncate and where not;

ii. if there are several spectral components, we'd always cut off a component at the wrong time (for the purists: except if the sampling interval can be chosen to be equal to the smallest common denominator of the single periods).

Example 4.12 (Truncation) See what happens for $N = 4$:

Without truncation error:

$$f(t) = \cos \omega t \qquad \text{with } \omega = \frac{2\pi}{4\Delta t}$$

With maximum truncation error:

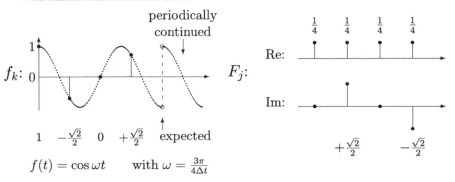

$$f(t) = \cos \omega t \qquad \text{with } \omega = \frac{3\pi}{4\Delta t}$$

$$W_4 = e^{i\pi/2}$$

$$F_0 = \frac{1}{4} \quad \text{(average)}$$

$$F_1 = \frac{1}{4}\left(1 + \left(-\frac{1}{\sqrt{2}}\right) \times \text{``6:00 a.m.''} + \left(+\frac{1}{\sqrt{2}}\right) \times \text{``12:00 noon''}\right)$$

$$= \frac{1}{4}\left(1 + \frac{i}{\sqrt{2}} + \frac{i}{\sqrt{2}}\right) = \frac{1}{4} + \frac{i}{2\sqrt{2}} \tag{4.40}$$

$$F_2 = \frac{1}{4}\left(1 + \left(-\frac{1}{\sqrt{2}}\right) \times \text{``9:00 a.m.''} + \left(+\frac{1}{\sqrt{2}}\right) \times \text{``9:00 p.m.''}\right)$$

$$= \frac{1}{4}\left(1 + \frac{1}{\sqrt{2}} - \frac{1}{\sqrt{2}}\right) = \frac{1}{4}$$

$$F_3 = F_1^*.$$

Two "strange findings":

i. Through truncation we suddenly got an imaginary part, in spite of using a cosine as input. But our function isn't *even* at all, because we continue using $f_N = -1$, instead of $f_N = f_0 = +1$, as we originally intended to do. This function contains an *even* and an *odd* portion (cf. Fig. 4.17).

ii. We really had expected a Fourier coefficient *between* half the Nyquist frequency and the Nyquist frequency, possibly spread evenly over F_1 and F_2, and not a constant, like we would have had to expect for the case of a δ-function as input: but we've precisely entered this as "even" input.

The "odd" input is a sine wave with amplitude $-1/\sqrt{2}$ and therefore results in an imaginary part of $F_1 = 1/2\sqrt{2}$; the intensity $-1/2\sqrt{2}$, split "between sisters", is to be found at F_3, the positive sign in front of $\mathrm{Im}\{F_1\}$ means *negative* amplitude (cf. the remarks in (4.14) about Bavarian clocks).

Instead of more discussions about truncation errors in the case of cosine inputs, we recall that $\omega = 0$ is a frequency "as good as any". So we want to discuss the discrete analog to the function $\frac{\sin x}{x}$, the Fourier transform of the "rectangular function". We use as input:

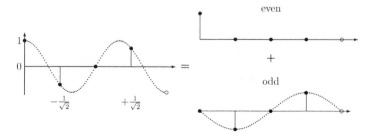

Fig. 4.17 Decomposition of the input into an even and an odd portion

$$f_k = \begin{cases} 1 \text{ for } 0 \le k \le M \\ 0 \text{ else} \\ 1 \text{ for } N - M \le k \le N - 1 \end{cases} \tag{4.41}$$

and stick with period N. This corresponds to a "rectangular window" of width $2M+1$ (M arbitrary, yet $< N/2$). Here, the half corresponding to negative times has been wrapped onto the right end of the interval. Please note, that we can't help but require an odd number of $f_k \ne 0$ to get an even function. An example with $N = 8$, $M = 2$ is shown in Fig. 4.18.

For general $M < N/2$ and N the Fourier transform is:

$$F_j = \frac{1}{N}\left(\sum_{k=0}^{M} W_N^{-kj} + \sum_{k=N-M}^{N-1} W_N^{-kj} \right)$$

$$= \frac{1}{N}\left(2\sum_{k=0}^{M} \cos(2\pi kj/N) - 1 \right).$$

The sum can be calculated using (1.53), which we came across when dealing with Dirichlet's integral kernel. We have:

$$\frac{1}{2} + \frac{1}{2} + \cos x + \cos 2x + \cdots + \cos Mx = \frac{1}{2} + \frac{\sin\left(M + \frac{1}{2}\right)x}{2\sin\frac{x}{2}}$$

$$\text{with } x = 2\pi j/N,$$

thus:

$$F_j = \frac{1}{N}\left(1 + \frac{\sin\left(M + \frac{1}{2}\right)\frac{2\pi j}{N}}{\sin\frac{2\pi j}{2N}} - 1 \right) = \frac{1}{N}\left(\frac{\sin\frac{2M+1}{N}\pi j}{\sin\frac{\pi j}{N}} \right) \tag{4.42}$$

$$\text{for } j = 0, \ldots, N - 1.$$

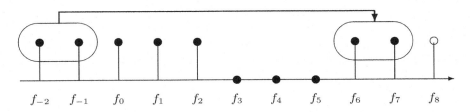

Fig. 4.18 "Rectangular" input using $N = 8$, $M = 2$

Fig. 4.19 Equation (4.42) (*points*); $\frac{2M+1}{N}\frac{\sin x}{x}$ with $x = \frac{2M+1}{N}\pi j$ (*thin line*)

This is the discrete version of the function $\frac{\sin x}{x}$ which we would get in the case of the continuous Fourier transformation (cf. Fig. 2.1 for our above example). Figure 4.19 shows the result for $N = 64$ and $M = 8$ in comparison to $\frac{\sin x}{x}$.

What happens at $j = 0$? There's a trick: j temporarily is treated like a continuous variable and l'Hospital's rule is applied:

$$F_0 = \frac{1}{N}\frac{\left(\dfrac{2M+1}{N}\right)\pi}{\pi/N} = \frac{2M+1}{N} \qquad \text{``average''}. \tag{4.43}$$

We had used $2M + 1$ series elements $f_k = 1$ as input. Only in this range the denominator can become 0.

Where are the zeros of the discrete Fourier transform of the discrete "rectangular window"? Funny, there is no F_j, that is exactly equal to 0, as $\frac{2M+1}{N}\pi j = l\pi$, $l = 1, 2, \ldots$ or $j = l\frac{N}{2M+1}$ and $j =$ integer can only be achieved for $l = 2M+1$, and then j already is outside the interval. Of course, for $M \gg 1$ we may approximately put $j \approx l\frac{N}{2M}$ and then get $2M - 1$ "quasi-zero transitions". This is different compared to the function $\frac{\sin x}{x}$, where there are real zeros. The oscillations around zero next to the central peak at $j = 0$ decay only very slowly; even worse, after $j = N/2$ the denominator starts getting smaller, and the oscillations increase again! Don't panic: in the right half of $\{F_j\}$ there is the mirror image of the left half! What's behind the difference to the function $\frac{\sin x}{x}$? It's the periodic continuation in the case of the discrete Fourier transformation! We transform a "comb" of "rectangular functions"! For $j \ll N$, i.e. far from the end of the interval, we get:

$$F_j = \frac{1}{N}\frac{\sin\dfrac{2M+1}{N}\pi j}{\pi j/N} = \frac{2M+1}{N}\frac{\sin x}{x} \qquad \text{with } x = \frac{2M+1}{N}\pi j, \tag{4.44}$$

and that's exactly what we'd have expected in the first place. In the extreme case of $M = N/2 - 1$ we get for $j \neq 0$ from (4.42):

$$ F_j = \frac{1}{N} \frac{\sin \dfrac{N-1}{N} \pi j}{\sin(\pi j / N)} = -\frac{1}{N} e^{i\pi j}, $$

which we can just manage to compensate by plugging the "hole" at $f_{N/2}$ (cf. Sect. 4.5, cosine transformation). Let's take a closer look at the extreme case of large N and large M (but with $2M \ll N$). In this limit we really get the same "zeros" as in function $\frac{\sin x}{x}$. Here we have a situation somewhat like the transition from the discrete to the continuous Fourier transformation (especially so if we only look at the Fourier coefficients F_j with $j \ll N$). Now we also understand why there are no sidelobes in the case of a discrete Fourier transformation of a cosine input without truncation errors and without zero-padding: the Fourier coefficients neighbouring the central peak are precisely where the zeros are. Then the Fourier transformation works like a—meanwhile obsolete—vibrating-reed frequency meter. This sort of instrument was used in earlier times to monitor the mains frequency of 50 cycles (60 cycles in the US and some other countries). Reeds with distinctive eigen-frequencies, for example 47, 48, 49, 50, 51, 52, 53 cycles, are activated using a mains-driven coil: only the reed with the proper eigen-frequency at the current mains-frequency will start vibrating, all others will keep quiet. These days, no energy supplier will get away with supplying 49 or 51 cycles, as this would cause all inexpensive (alarm)clocks (without quartz-control) to get out of sync. What's true for the frequency $\omega = 0$, of course also is true for all other frequencies $\omega \neq 0$, according to the Convolution Theorem. This means that we can only get a consistent line profile of a spectral line that doesn't depend on truncation errors if we use zero-padding, and make it plenty of zeros.

So here is the *1st recommendation*:

Many zeros are good! N very big; $2M \ll N$.

The economy and politics also obey this rule.

Since available Fourier transformation programs do not offer zero-padding you have to do it yourself.

2nd recommendation:

Choose your sampling-interval Δt fine enough, so that your Nyquist frequency is always substantially higher than the expected spectral intensity, meaning, you need F_j only for $j \ll N$. This should get rid of the consequences of the periodic continuation approximately!

In Chap. 3 we quite extensively discussed continuous window functions. A very good presentation of window functions in the case of the discrete Fourier transformation can be found in Harris [9]. We're happy to know, however, that we may transfer

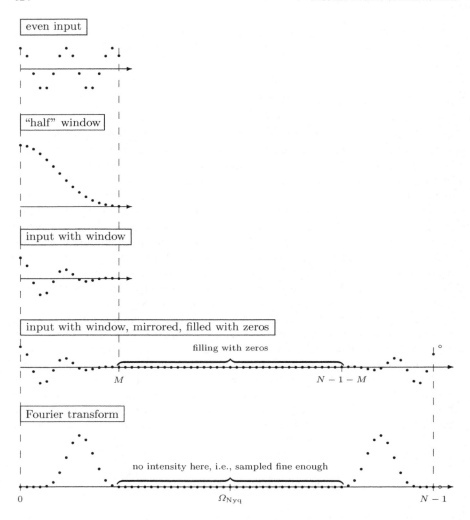

Fig. 4.20 "Cooking recipe" for the Fourier transformation for an even input; in case of an odd input invert the mirror image

all the properties of a continuous window function to the discrete Fourier transformation straight away, if, by using enough zeros for padding and using the low-frequency portion of the Fourier series, we aim for the limes discrete → continuous.

So here comes the *3rd recommendation*:

Do use window functions!

These three recommendations are illustrated in Fig. 4.20 in an easy-to-remember way. If you know that the input is even or odd, respectively, data mirroring is always recommended. Should you have zero-padded your data you must window your data

yourself and eventually use mirroring before clicking the FFT button. Do not use offered window functions, they would extend over the zero-padded region as well. Then use the "rectangular" window, i.e. no further window.

If the input is neither even nor odd, you can force the input to become even or odd, respectively, provided all spectral components have the same phase. The situation is more complicated if the input contains even and odd components, i.e. the spectral components have different phases. If these components are well separated you can shift the phase for each component individually. If these components are not well separated use the full window function, i.e. don't mirror the data, than zero-padd and Fourier transform. Now, the real and the imaginary part depend on where you zero-padd: at the beginning, at the end, or both. In this case the power representation is recommended.

In spite of the fact that today's fast PCs won't have a problem transforming very big data sets any more, the application of the Fourier transformation got a huge boost from the "Fast Fourier transformation" algorithm by Cooley and Tukey, an algorithm that doesn't grow with N^2 calculations but only $N \ln N$.

We'll have a closer look at this algorithm in the next section.

4.7 Fast Fourier Transformation (FFT)

Cooley and Tukey started out from the simple question: what is the Fourier transform of a series of numbers with only **one** real number ($N = 1$)? There are at least 3 answers:

i. *Accountant*:
   From (4.12) with $N = 1$ follows:

$$F_0 = \tfrac{1}{1} f_0 W_1^{-0} = f_0. \tag{4.45}$$

ii. *Economist*:
    From (4.31) (Parseval's theorem) follows:

$$|F_0|^2 = \tfrac{1}{1} |(f_0)|^2. \tag{4.46}$$

Using the services of *someone into law*: f_0 is real and even, which leads to $F_0 = \pm f_0$, and as F_0 is also to be equal to the average of the series of numbers, there's no chance of getting a minus sign.
(A *layperson* would have done without all this lead-in talk!)

iii. *Philosopher*:
     We know that the Fourier transform of a δ-function results in a constant and vice versa. How do we represent a constant in the world of 1-term series? By using the number f_0. How do we represent in this world a δ-function? By using this number f_0. So in this world there's no difference any more between a constant and a δ-function. Result: f_0 is its own Fourier transform.

This finding, together with the trick to achieve $N = 1$ by smartly halving the input again and again (that's why we have to stipulate: $N = 2^p$, p integer), (almost) saves us the Fourier transformation. For this purpose, let's first have a look at the first subdivision. We'll assume as given: $\{f_k\}$ with $N = 2^p$. This series will get cut up in a way that one subseries will only contain the even elements and the other subseries only the odd elements of $\{f_k\}$:

$$\begin{aligned} \{f_{1,k}\} &= \{f_{2k}\} & k &= 0, 1, \ldots, M-1, \\ \{f_{2,k}\} &= \{f_{2k+1}\} & M &= N/2. \end{aligned} \tag{4.47}$$

Both subseries are periodic in M.

Proof (Periodicity in M)

$$f_{1,k+M} = f_{2k+2M} = f_{2k} = f_{1,k}$$
$$\text{because of } 2M = N \text{ and } f \text{ periodic in } N.$$

Analogously for $f_{2,k}$. $\square$

The respective Fourier transforms are:

$$\begin{aligned} F_{1,j} &= \frac{1}{M} \sum_{k=0}^{M-1} f_{1,k} W_M^{-kj}, \\ F_{2,j} &= \frac{1}{M} \sum_{k=0}^{M-1} f_{2,k} W_M^{-kj}, \end{aligned} \qquad j = 0, \ldots, M-1. \tag{4.48}$$

The Fourier transform of the original series is:

$$\begin{aligned} F_j &= \frac{1}{N} \sum_{k=0}^{N-1} f_k W_M^{-kj} \\ &= \frac{1}{N} \sum_{k=0}^{M-1} f_{2k} W_N^{-2kj} + \frac{1}{N} \sum_{k=0}^{M-1} f_{2k+1} W_N^{-(2k+1)j} \\ &= \frac{1}{N} \sum_{k=0}^{M-1} f_{1,k} W_M^{-kj} + \frac{W_N^{-j}}{N} \sum_{k=0}^{M-1} f_{2,k} W_M^{-kj}, \qquad j = 0, \ldots, N-1. \end{aligned} \tag{4.49}$$

In our last step we used:

$$\begin{aligned} W_N^{-2kj} &= e^{-2 \times 2\pi i kj / N} = e^{-2\pi i kj/(N/2)} = W_M^{-kj}, \\ W_N^{-(2k+1)j} &= e^{-2\pi i (2k+1)j/N} = W_M^{-kj} W_N^{-j}. \end{aligned}$$

Together we get:

$$F_j = \tfrac{1}{2}F_{1,j} + \tfrac{1}{2}W_N^{-j}F_{2,j}, \qquad j = 0, \dots, N-1,$$

or better:

$$
\begin{aligned}
F_j &= \tfrac{1}{2}(F_{1,j} + F_{2,j}W_N^{-j}),\\
F_{j+M} &= \tfrac{1}{2}(F_{1,j} - F_{2,j}W_N^{-j}), \qquad j = 0, \dots, M-1.
\end{aligned}
\tag{4.50}
$$

Please note that in (4.50) we allowed j to run from 0 to $M-1$ only. In the second line in front of $F_{2,j}$ we have used:

$$
\begin{aligned}
W_N^{-(j+M)} &= W_N^{-j}W_N^{-M} = W_N^{-j}W_N^{-N/2} = W_N^{-j}e^{-2\pi i \frac{N}{2}/N}\\
&= W_N^{-j}e^{-i\pi} = -W_N^{-j}.
\end{aligned}
\tag{4.51}
$$

This "decimation in time" can be repeated until we finally end up with 1-term series whose Fourier transforms are identical to the input number, as we know. The conventional Fourier transformation requires N^2 calculations, whereas here we only need $pN = N \ln N$.

Example 4.13 ("Saw-tooth" with $N = 2$)

$$f_0 = 0, \qquad f_1 = 1.$$

Normal Fourier transformation:

$$
\begin{aligned}
W_2 &= e^{i\pi} = -1\\
F_0 &= \frac{1}{2}(0+1) = \frac{1}{2}\\
F_1 &= \frac{1}{2}\left(0 + 1 \times W_2^{-1}\right) = -\frac{1}{2}.
\end{aligned}
\tag{4.52}
$$

Fast Fourier transformation:

$$
\begin{aligned}
f_{1,0} &= 0 \text{ even part} \;\to\; F_{1,0} = 0\\
f_{2,0} &= 1 \text{ odd part} \;\to\; F_{2,0} = 1, \qquad M = 1.
\end{aligned}
\tag{4.53}
$$

From formula (4.50) we get:

$$
\begin{aligned}
F_0 &= \frac{1}{2}\left(F_{1,0} + F_{2,0}\,\underbrace{W_2^0}_{=1}\right) = \frac{1}{2}\\
F_1 &= \frac{1}{2}\left(F_{1,0} - F_{2,0}W_2^0\right) = -\frac{1}{2}.
\end{aligned}
\tag{4.54}
$$

This didn't really save all that much work so far.

Example 4.14 ("Saw-tooth" with N = 4)

$$f_0 = 0$$
$$f_1 = 1$$
$$f_2 = 2$$
$$f_3 = 3.$$

The normal Fourier transformation gives us:

$$W_4 = e^{2\pi i/4} = e^{\pi i/2} = i$$

$$F_0 = \frac{1}{4}(0 + 1 + 2 + 3) = \frac{3}{2} \quad \text{"average"}$$

$$F_1 = \frac{1}{4}\left(W_4^{-1} + 2W_4^{-2} + 3W_4^{-3}\right) = \frac{1}{4}\left(\frac{1}{i} + \frac{2}{-1} + \frac{3}{-i}\right) = -\frac{1}{2} + \frac{i}{2} \quad (4.55)$$

$$F_2 = \frac{1}{4}\left(W_4^{-2} + 2W_4^{-4} + 3W_4^{-6}\right) = \frac{1}{4}(-1 + 2 - 3) = -\frac{1}{2}$$

$$F_3 = \frac{1}{4}\left(W_4^{-3} + 2W_4^{-6} + 3W_4^{-9}\right) = \frac{1}{4}\left(-\frac{1}{i} - 2 + \frac{3}{i}\right) = -\frac{1}{2} - \frac{i}{2}.$$

This time we're not using the trick with the clock, yet another playful approach. We can skillfully subdivide the input and thus get the Fourier transform straight away (cf. Fig. 4.21).

Using 2 subdivisions, the Fast Fourier transformation gives us:
1st subdivision:

$$\begin{array}{ll} N = 4 & \{f_1\} = \{0, 2\} \text{ even,} \\ M = 2 & \{f_2\} = \{1, 3\} \text{ odd.} \end{array} \quad (4.56)$$

2nd subdivision ($M' = 1$):

$$f_{11} = 0 \text{ even} \equiv F_{1,1,0},$$
$$f_{12} = 2 \text{ odd} \equiv F_{1,2,0},$$
$$f_{21} = 1 \text{ even} \equiv F_{2,1,0},$$
$$f_{22} = 3 \text{ odd} \equiv F_{2,2,0}.$$

Using (4.50) this results in ($j = 0, M' = 1$):

$$\begin{array}{cc} \text{upper part} & \text{lower part} \end{array}$$

$$F_{1,k} = \left\{\frac{1}{2}F_{1,1,0} + \frac{1}{2}F_{1,2,0}, \frac{1}{2}F_{1,1,0} - \frac{1}{2}F_{1,2,0}\right\} = \{1, -1\},$$

$$F_{2,k} = \left\{\frac{1}{2}F_{2,1,0} + \frac{1}{2}F_{2,2,0}, \frac{1}{2}F_{2,1,0} - \frac{1}{2}F_{2,2,0}\right\} = \{2, -1\}$$

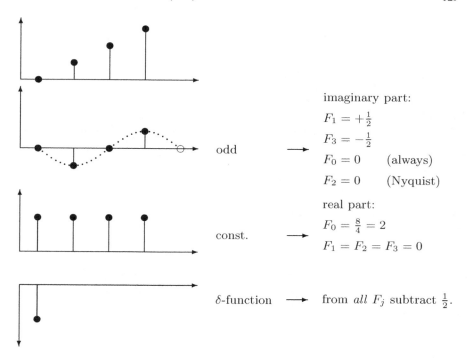

imaginary part:

$F_1 = +\frac{1}{2}$

$F_3 = -\frac{1}{2}$

$F_0 = 0 \qquad$ (always)

$F_2 = 0 \qquad$ (Nyquist)

real part:

$F_0 = \frac{8}{4} = 2$

$F_1 = F_2 = F_3 = 0$

from *all* F_j subtract $\frac{1}{2}$.

Fig. 4.21 Decomposition of the saw-tooth into an odd part, constant plus δ-function. Add up the F_k, and compare the result with (4.55)

and finally, using (4.50) once again:

upper part $\begin{cases} F_0 = \dfrac{1}{2}(F_{1,0} + F_{2,0}) = \dfrac{3}{2}, \\[2mm] F_1 = \dfrac{1}{2}\left(F_{1,1} + F_{2,1}W_4^{-1}\right) = \dfrac{1}{2}\left(-1 + (-1)\times\dfrac{1}{i}\right) = -\dfrac{1}{2} + \dfrac{i}{2}, \end{cases}$

lower part $\begin{cases} F_2 = \dfrac{1}{2}(F_{1,0} - F_{2,0}) = -\dfrac{1}{2}, \\[2mm] F_3 = \dfrac{1}{2}\left(F_{1,1} - F_{2,1}W_4^{-1}\right) = \dfrac{1}{2}\left(-1 - (-1)\times\dfrac{1}{i}\right) = -\dfrac{1}{2} - \dfrac{i}{2}. \end{cases}$

We can represent the calculations we've just done in the following diagram, where we've left out the factors $1/2$ per subdivision—they can be taken into account at the end when calculating the F_j (Fig. 4.22).

Here $\nearrow \oplus$ means add and $\searrow \ominus$ subtract and W_4^{-j} multiply with weight W_4^{-j}. This subdivision is called "decimation in time"; the scheme:

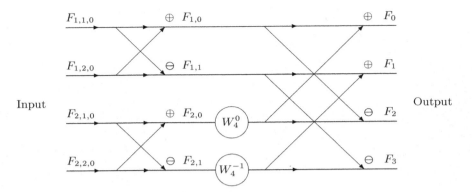

Fig. 4.22 Flow-diagram for the FFT with $N = 4$

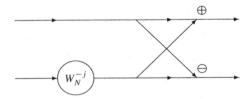

is called "butterfly scheme", which, for example, is used as a building-block in hardware Fourier analysers. Figure 4.23 illustrates the scheme for $N = 16$.

Those in the know will have found that the input is not required in the normal order $f_0 \ldots f_N$, but in bit-reversed order (arabic from right to left).

Example 4.15 (Bit-reversal for $N = 16$)

k	binary	reversed	results in k'
0	0000	0000	0
1	0001	1000	8
2	0010	0100	4
3	0011	1100	12
4	0100	0010	2
5	0101	1010	10
6	0110	0110	6
7	0111	1110	14
8	1000	0001	1
9	1001	1001	9
10	1010	0101	5
11	1011	1101	13
12	1100	0011	3
13	1101	1011	11
14	1110	0111	7
15	1111	1111	15

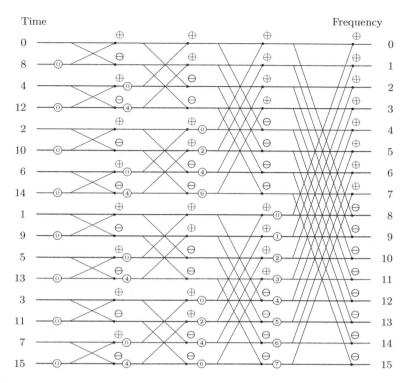

Fig. 4.23 Decimation in Time with $N = 16$

Computers have no problem with this bit-reversal.

At the end, let's have a look at a simple example:

$$\text{with for example } \textcircled{7} = \left(W_{16}^{-7}\right).$$

Example 4.16 (Half Nyquist frequency)

$$f_k = \cos(\pi k/2), \quad k = 0, \ldots, 15, \quad \text{i.e.}$$
$$f_0 = f_4 = f_8 = f_{12} = 1,$$
$$f_2 = f_6 = f_{10} = f_{14} = -1,$$
$$\text{all odd ones are 0.}$$

The bit-reversal orders the input in such a way that we get zeros in the lower half (cf. Fig. 4.24). If both inputs of the "butterfly scheme" are 0, i.e. we surely get 0 at the output, we do not show the add-/subtract-crosses. The intermediate results of the required calculations are quoted. The weights $W_{16}^0 = 1$ are not quoted for the sake of clarity. Other powers do not show up in this example. You see, the input is progressively "compressed" in four steps. Finally, we find a number 8 at negative and

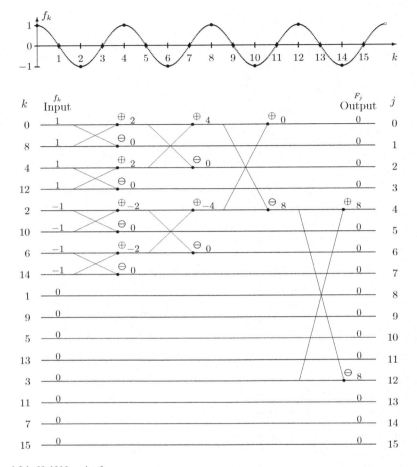

Fig. 4.24 Half Nyquist frequency

positive half Nyquist frequency each which we are allowed to add and subsequently have to divide by 16 which finally yields the amplitude of the cosine input, i.e. 1.

Playground

4.1 Correlated
What is the cross correlation of a series $\{f_k\}$ with a constant series $\{g_k\}$? Sketch the procedure with Fourier transforms!

4.2 No Common Ground
Given is the series $\{f_k\} = \{1, 0, -1, 0\}$ and the series $\{g_k\} = \{1, -1, 1, -1\}$.
   Calculate the cross correlation of the two series.

4.3 Brotherly

Calculate the cross correlation of $\{f_k\} = \{1, 0, 1, 0\}$ and $\{g_k\} = \{1, -1, 1, -1\}$, use the Convolution Theorem.

4.4 Autocorrelated

Given is the series $\{f_k\} = \{0, 1, 2, 3, 2, 1\}$, $N = 6$.

   Calculate its autocorrelation function. Check your results by calculating the Fourier transform of f_k and of $f_k \otimes f_k$.

4.5 Shifting around

Given the following input series (see Fig. 4.25):

$$f_0 = 1, \qquad f_k = 0 \quad \text{for } k = 1, \ldots, N - 1.$$

a. Is the series even, odd, or mixed?
b. What is the Fourier transform of this series?
c. The discrete "δ-function" now gets shifted to f_1 (Fig. 4.26). Is the series even, odd, or mixed?
d. What do we get for $|F_j|^2$?

4.6 Pure Noise

Given the random input series containing numbers between -0.5 and 0.5.

a. What does the Fourier transform of a random series look like (see Fig. 4.27)?
b. How big is the noise power of the random series, defined as:

$$\sum_{j=0}^{N-1} |F_j|^2? \tag{4.57}$$

Compare the result in the limiting case of $N \to \infty$ to the signal power of the input $0.5 \cos \omega t$.

Fig. 4.25 Input-signal with a δ-shaped pulse at $k = 0$

Fig. 4.26 Input-signal with a δ-shaped pulse at $k = 1$

Fig. 4.27 Random series

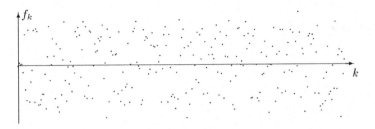

Fig. 4.28 Input function according to (4.58)

Fig. 4.29 Theoretical pattern ("toothbrush") that's to be located in the data set

4.7 Pattern Recognition

Given a sum of cosine functions as input, with plenty of superimposed noise
(Fig. 4.28):

$$f_k = \cos\frac{5\pi k}{32} + \cos\frac{7\pi k}{32} + \cos\frac{9\pi k}{32} + 15(\text{RND} - 0.5) \qquad (4.58)$$
$$\text{for } k = 0, \ldots, 255,$$

where RND is a random number[1] between 0 and 1.

How do you look for the pattern Fig. 4.29 that's buried in the noise, if it represents
the three cosine functions with the frequency ratios $\omega_1 : \omega_2 : \omega_3 = 5 : 7 : 9$?

4.8 Go on the Ramp (for Gourmets only)

Given the input series:

$$f_k = k \text{ for } k = 0, 1, \ldots, N - 1.$$

---

[1]Programming languages such as, for example Turbo-Pascal, C, Fortran, … feature random gener-
ators that can be called as functions. Their efficiency varies considerably.

Is this series even, odd, or mixed? Calculate the real and imaginary part of its Fourier transform. Check your results using Parseval's theorem. Derive the results for $\sum_{j=1}^{N-1} 1/\sin^2(\pi j/N)$ and $\sum_{j=1}^{N-1} \cot^2(\pi j/N)$.

4.9 Transcendental (for Gourmets only)

Given the input series (with N even!):

$$f_k = \begin{cases} k & \text{for } k = 0, 1, \ldots, \frac{N}{2} - 1 \\ N - k & \text{for } k = \frac{N}{2}, \frac{N}{2} + 1, \ldots, N - 1 \end{cases}. \tag{4.59}$$

Is the series even, odd, or mixed? Calculate its Fourier transform. The double-sided ramp is a high-pass filter (cf. Sect. 5.2), which immediately becomes clear considering the periodic continuation. Use Parseval's theorem to derive the result for $\sum_{k=1}^{N/2} 1/\sin^4(\pi(2k-1)/N)$. Use the fact that a high-pass does not transfer a constant in order to derive the result for $\sum_{k=1}^{N/2} 1/\sin^2(\pi(2k-1)/N)$.

Chapter 5
Filter Effect in Digital Data Processing

Abstract This chapter deals with filter effects in digital data processing. For this purpose the transfer function is introduced. Simple filters like high-pass, low-pass, band-pass, and notch are discussed. The effects of data shifting, data compression as well as differentiation and integration of discrete data are shown.

In this chapter we'll discuss only very simple procedures, such as smoothing of data, shifting of data using linear interpolation, compression of data, differentiating data and integrating them, and while doing that, describe the filter effect—something that's often not even known to our subconscience. For this purpose, the concept of the transfer function comes in handy.

5.1 Transfer Function

We'll take as given a "recipe" according to which the output $y(t)$ is made up of a linear combination of $f(t)$ including derivatives and integrals:

$$\underbrace{y(t)}_{\text{"output"}} = \sum_{j=-k}^{+k} a_j \underbrace{f^{[j]}(t)}_{\text{"input"}} \tag{5.1}$$

$$\text{with } f^{[j]} = \frac{d^j f(t)}{dt^j} \quad \text{(negative } j \text{ means integration).}$$

This rule is linear and stationary, as a shift along the time axis in the input results in the same shift along the time axis in the output.

When we Fourier-transform (5.1) we get with (2.57):

$$Y(\omega) = \sum_{j=-k}^{+k} a_j \text{FT}\left(f^{[j]}(t)\right) = \sum_{j=-k}^{+k} a_j (i\omega)^j F(\omega) \tag{5.2}$$

© Springer International Publishing Switzerland 2015
T. Butz, *Fourier Transformation for Pedestrians*, Undergraduate Lecture Notes in Physics, DOI 10.1007/978-3-319-16985-9_5

or:

$$Y(\omega) = H(\omega)F(\omega)$$

with the transfer function $H(\omega) = \sum_{j=-k}^{+k} a_j (\mathrm{i}\omega)^j .$ \hfill (5.3)

When looking at (5.3), we immediately think of the Convolution Theorem. According to this, we may interpret $H(\omega)$ as the Fourier transform of the output $y(t)$ using δ-shaped input (that's $F(\omega) = 1$). So weighted with this transfer function, $F(\omega)$ is translated into the output $Y(\omega)$. In the frequency domain, we can easily filter if we choose an adequate $H(\omega)$. Here, however, we want to work in the time domain.

Now we'll get into number series. Please note that we'll get derivatives only over differences and integrals only over sums of single discrete numbers. Therefore we'll have to widen the definition (5.1) by including *non-stationary* parts. The operator V^l means shift by l:

$$V^l y_k \equiv y_{k+l}. \hfill (5.4)$$

This allows us to state the discrete version of (5.1) as follows:

$$\underbrace{y_k}_{\text{"output"}} = \sum_{l=-L}^{+L} a_l \underbrace{V^l f_k}_{\text{"input"}} . \hfill (5.5)$$

Here, positive l stand for *later* input samples, and negative l for *earlier* input samples. With positive l, we can't process a data-stream sequentially in "real-time", we first have to buffer L samples, for example in a shift-register, which often is called a FIFO (first in, first out). These algorithms are called acausal. The Fourier transformation is an example for an acausal algorithm.

The discrete Fourier transformation of (5.5) is:

$$Y_j = \sum_{l=-L}^{+L} a_l \mathrm{FT}\left(V^l f_k\right) = \sum_{l=-L}^{+L} a_l \frac{1}{N} \sum_{k=0}^{N-1} f_{k+l} W_N^{-kj}$$

$$= \sum_{l=-L}^{+L} a_l \frac{1}{N} \sum_{k'=l}^{N-1+l} f_{k'} W_N^{-k'j} W_N^{+lj}$$

$$= \sum_{l=-L}^{+L} a_l W_N^{+lj} F_j = H_j F_j.$$

$$Y_j = H_j F_j$$

$$\text{with } H_j = \sum_{l=-L}^{+L} a_l W_N^{+lj} = \sum_{l=-L}^{+L} a_l e^{i\omega_j l \Delta t} \text{ and } \omega_j = 2\pi j/(N \Delta t). \tag{5.6}$$

Using this transfer function, which we assume to be continuous *out of pure convenience*,[1] that's $H(\omega) = \sum_{l=-L}^{+L} a_l e^{i\omega l \Delta t}$, it's easy to understand the "filter effects" of the previously defined operations.

5.2 Low-Pass, High-Pass, Band-Pass, Notch Filter

First we'll look into the filter effect when smoothing data. A simple 2-point algorithm for data-smoothing would be, for example:

$$y_k = \frac{1}{2}(f_k + f_{k+1}) \tag{5.7}$$
$$\text{with} \quad a_0 = \frac{1}{2}, \quad a_1 = \frac{1}{2}.$$

This gives us the transfer function:

$$H(\omega) = \frac{1}{2}\left(1 + e^{i\omega \Delta t}\right). \tag{5.8}$$

$$|H(\omega)|^2 = \frac{1}{4}(1 + e^{i\omega \Delta t})(1 + e^{-i\omega \Delta t}) = \frac{1}{2} + \frac{1}{2}\cos\omega t = \cos^2\frac{\omega \Delta t}{2}$$

and finally:

$$|H(\omega)| = \cos\frac{\omega \Delta t}{2}.$$

Figure 5.1 shows $|H(\omega)|$.

This has the unpleasant effect that a real input results in a complex output. This, of course, is due to our implicitly introduced "phase shift" by $\Delta t/2$.

It looks like the following 3-point algorithm will do better:

$$y_k = \frac{1}{3}(f_{k-1} + f_k + f_{k+1}) \tag{5.9}$$
$$\text{with} \quad a_{-1} = \frac{1}{3}, \quad a_0 = \frac{1}{3}, \quad a_1 = \frac{1}{3}.$$

---

[1] We can always choose N to be large, so j is very dense.

Fig. 5.1 Modulus of the
transfer function for the
smoothing-algorithm of (5.7)

This gives us:

$$H(\omega) = \frac{1}{3}\left(e^{-i\omega\Delta t} + 1 + e^{+i\omega\Delta t}\right) = \frac{1}{3}(1 + 2\cos\omega\Delta t). \tag{5.10}$$

Figure 5.2 shows $H(\omega)$ and the problem that for $\omega = 2\pi/3\Delta t$ there is a zero, meaning that this frequency will not get transferred at all. This frequency is $(2/3)\Omega_{Nyq}$. Above that, even the sign changes. This algorithm is not consistent and therefore should not be used.

The "correct" smoothing-algorithm is as follows:

$$y_k = \frac{1}{4}\left(f_{k-1} + 2f_k + f_{k+1}\right) \qquad \text{low-pass} \tag{5.11}$$

$$\text{with} \quad a_{-1} = +\frac{1}{4}, \quad a_0 = +\frac{1}{2}, \quad a_1 = +\frac{1}{4}.$$

The transfer function now reads:

$$H(\omega) = \frac{1}{4}\left(e^{-i\omega\Delta t} + 2 + e^{+i\omega\Delta t}\right)$$
$$= \frac{1}{4}(2 + 2\cos\omega\Delta t) = \cos^2\frac{\omega\Delta t}{2}. \tag{5.12}$$

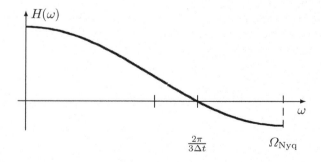

Fig. 5.2 Transfer function for the 3-point smoothing-algorithm as of (5.9)

Fig. 5.3 Transfer function for the low-pass

Figure 5.3 shows $H(\omega)$: there are no zeros, the sign doesn't change. Comparing this to (5.8) and Fig. 5.1, it's obvious that the filter effect now is bigger: $\cos^2(\omega\Delta t/2)$ instead of $\cos(\omega\Delta t/2)$ for $|H(\omega)|$.

Using half the Nyquist frequency we get:

$$H(\Omega_{\text{Nyq}}/2) = \cos^2\frac{\pi}{4} = \frac{1}{2}.$$

Therefore our smoothing-algorithm is a low-pass filter, which, admittedly, doesn't have a "very steep edge", and which, at $\omega = \Omega_{\text{Nyq}}/2$, will let only half the amount pass. So at $\omega = \Omega_{\text{Nyq}}/2$ we have -3 dB attenuation.

If our data is corrupted by low-frequency artefacts (for example slow drifts), we'd like to use a high-pass filter. Here's how we design it:

$$H(\omega) = 1 - \cos^2\frac{\omega\Delta t}{2} = \sin^2\frac{\omega\Delta t}{2}$$
$$= \frac{1}{2}(1 - \cos\omega\Delta t)$$
$$= \frac{1}{2}\left(1 - \frac{1}{2}e^{-i\omega\Delta t} - \frac{1}{2}e^{+i\omega\Delta t}\right). \qquad (5.13)$$

So we have: $a_{-1} = -1/4$, $a_0 = +1/2$, $a_1 = -1/4$, and the algorithm is:

$$y_k = \frac{1}{4}\left(-f_{k-1} + 2f_k - f_{k+1}\right) \qquad \text{high-pass.} \qquad (5.14)$$

From (5.14) we realise straight away: a constant as input will not get through because the sum of the coefficients a_i is zero.

Figure 5.4 shows $H(\omega)$. Here, too, we can see that at $\omega = \Omega_{\text{Nyq}}/2$ half the amount will get through only. The experts talk of -3 dB attenuation at $\omega = \Omega_{\text{Nyq}}/2$. We discussed in Example 4.14 the "saw-tooth". In the frequency domain this is a high-pass, too! In a certain image reconstruction algorithm from many projections taken at different angles, as required in tomography, exactly such high-pass filters are in use. They are called ramp filters. They naturally show up when transforming from cartesian to cylinder coordinates. We shall discuss this algorithm, called "backprojection of filtered projections", in more detail in Chap. 7.

Fig. 5.4 Transfer function for the high-pass

Fig. 5.5 Transfer function of the band-pass

If we want to suppress very low as well as very high frequencies, we need a band-pass. For simplicity's sake we take the product of the previously described low- and high-pass (cf. Fig. 5.5):

$$H(\omega) = \cos^2 \frac{\omega \Delta t}{2} \sin^2 \frac{\omega \Delta t}{2} = \left(\frac{1}{2} \sin \omega \Delta t \right)^2$$

$$= \frac{1}{4} \sin^2 \omega \Delta t = \frac{1}{4} \frac{1}{2} (1 - \cos 2\omega \Delta t)$$

$$= \frac{1}{8} \left(1 - \frac{1}{2} e^{-2i\omega \Delta t} - \frac{1}{2} e^{+2i\omega \Delta t} \right). \tag{5.15}$$

So we have $a_{-2} = -1/16$, $a_0 = +1/8$, $a_{+2} = -1/16$ and:

$$f_k = \frac{1}{16} (-f_{k-2} + 2f_k - f_{k+2}) \qquad \text{band-pass.} \tag{5.16}$$

Now, at $\omega = \Omega_{\text{Nyq}}/2$ we have $H(\Omega_{\text{Nyq}}/2) = 1/4$, that's -6 dB attenuation. If we choose to set the complement of the band-pass to 1:

$$H(\omega) = 1 - \left(\frac{1}{2} \sin \omega \Delta t \right)^2, \tag{5.17}$$

we'll get a notch filter that suppresses frequencies around $\omega = \Omega_{\text{Nyq}}/2$, yet lets all others pass (cf. Fig. 5.6).

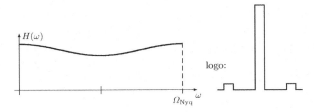

Fig. 5.6 Transfer function of the notch filter

$H(\omega)$ can be transformed to:

$$H(\omega) = 1 - \frac{1}{8} + \frac{1}{16}e^{2i\omega\Delta t} + \frac{1}{16}e^{-2i\omega\Delta t} \tag{5.18}$$

$$\text{with} \quad a_{-2} = +1/16, \quad a_0 = +7/8, \quad a_{+2} = +1/16$$

$$\text{and} \quad y_k = \frac{1}{16}\left(f_{k-2} + 14f_k + f_{k+2}\right) \qquad \text{notch filter.} \tag{5.19}$$

The suppression at half the Nyquist frequency, however, isn't exactly impressive: only a factor of $3/4$ or -1.25 dB.

Figure 5.7 gives an overview/recaps all the filters we've covered.

How can we build better notch filters? How can we set the cut-off frequency? How can we set the edge steepness? Linear, non-recursive filters won't do the job. Therefore we'll have to look at *recursive* filters, where part of the output is fed back as input. In RF-engineering this is called feedback. Live TV-shows with viewers calling in on their phones know what (acoustic) feedback is: it goes from your phone's mouthpiece via plenty of wire (copper or fibre) and various electronics to the studio's loudspeakers, and from there on to the microphone, the transmitter and back to your TV-set (maybe using a satellite for good measure) and on to your phone's handset. Quite an elaborate set-up, isn't it. No wonder we can have lots of fun letting rip on talkshows using this kind of feedback! Video-experts may use their cameras to achieve optical feedback by pointing it at the TV-screen that happens to show exactly this camera and so on. (This is the modern, yet chaos-inducing, version of the principle of the never-ending mirroring, using two mirrors opposite each other,

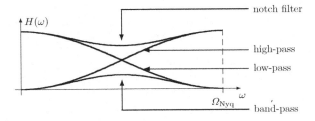

Fig. 5.7 Overview of the transfer functions of the various filters

like, for example, in the Mirror Hall of the Castle of Linderhof, a tourist attraction, the "little brother" of the castle Neuschwanstein of King Ludwig II. of Bavaria.)

It's not appropriate to discuss digital filters in depth here. We'll only look at a small example to glean the principles of a low-pass with a recursive algorithm. The algorithm may be formulated in a general manner as follows:

$$
y_k = \sum_{l=-L}^{L} a_l V^l f_k - \sum_{\substack{m=-M \\ m \neq 0}}^{M} b_m V^m y_k \tag{5.20}
$$

with the definition: $V^l f_k = f_{k+l}$ (as above). We arbitrarily chose the sign in front of the second sum to be negative; and for the same reason, we excluded $m = 0$ from the sum. Both moves will prove to be very useful shortly.

For negative m the *previous* output is fed back to the right hand side of (5.20), for the calculation of the new output: the algorithm is *causal*. For positive m the *subsequent* output is fed back for the calculation of the new output: the algorithm is *acausal*. Possible work-around: input and output are pushed into memory (register) and kept in intermediate storage as long as M is big.

We may transform (5.20) into:

$$
\sum_{m=-M}^{M} b_m V^m y_k = \sum_{l=-L}^{L} a_l V^l f_k. \tag{5.21}
$$

The Fourier transform of (5.21) may be rewritten, like in (5.6) (with $b_0 = 1$):

$$
B_j Y_j = A_j F_j \tag{5.22}
$$

$$
\text{with } B_j = \sum_{m=-M}^{M} b_m W_N^{+mj} \quad \text{and} \quad A_j = \sum_{l=-L}^{L} a_l W_N^{+lj}.
$$

So the output is $Y_j = \frac{A_j}{B_j} F_j$, and we may define the new transfer function as:

$$
H_j = \frac{A_j}{B_j} \quad \text{or} \quad H(\omega) = \frac{A(\omega)}{B(\omega)}. \tag{5.23}
$$

Using feedback we may, via zeros in the denominator, create poles in $H(\omega)$, or better, using somewhat less feedback, create *resonance enhancement*.

Example 5.1 (*Feedback*) Let's take our low-pass from (5.12) with 50 % feedback of the previous output:

$$y_k = \frac{1}{2}y_{k-1} + \frac{1}{4}(f_{k-1} + 2f_k + f_{k+1}) \quad \text{or}$$

$$\left(1 - \frac{1}{2}V^{-1}\right)y_k = \frac{1}{4}\left(V^{-1} + 2 + V^{+1}\right)f_k. \tag{5.24}$$

This results in:

$$H(\omega) = \frac{\cos^2(\omega\Delta t/2)}{1 - \frac{1}{2}e^{-i\omega\Delta t}}. \tag{5.25}$$

If we don't care about the phase shift, caused by the feedback, we're only interested in:

$$|H(\omega)| = \frac{\cos^2(\omega\Delta t/2)}{\sqrt{\left(1 - \frac{1}{2}\cos\omega\Delta t\right)^2 + \left(\frac{1}{2}\sin\omega\Delta t\right)^2}} = \frac{\cos^2(\omega\Delta t/2)}{\sqrt{\frac{5}{4} - \cos\omega\Delta t}}. \tag{5.26}$$

The *resonance enhancement* at $\omega = 0$ is 2, $|H(\omega)|$ is shown in Fig. 5.8, together with the non-recursive low-pass from (5.12). We can clearly see that the edge steepness got better. If we'd fed back 100 % instead of 50 % in (5.24), a single short input would have been enough to keep the output "high" for good; the filter would have been unstable. In our case, it decays like a geometric series once the input has been taken off.

Here we've already taken the first step into the highly interesting field of filters in the time domain. If you want to know more about it, have a look at, for example, "Numerical Recipes" [7] and the material quoted there. But don't forget that filters in the frequency domain are much easier to handle because of the Convolution Theorem. We shall discuss an interesting filter in more detail in Chap. 6.

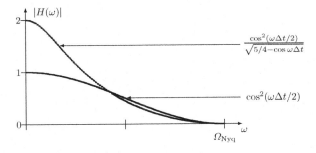

Fig. 5.8 Transfer function for the low-pass (5.12) and the filter with feedback (5.26)

5.3 Shifting Data

Let's assume you have a data set and you want to shift it a fraction d of the sampling interval Δt, say, for simplicity's sake, using linear interpolation. So you'd rather have started sampling d later, yet won't (or can't) repeat the measurements. Then you should use the following algorithm:

$$y_k = (1 - d) f_k + d f_{k+1}, \ 0 < d < 1 \quad \text{``shifting with linear interpolation''}. \tag{5.27}$$

The corresponding transfer function reads:

$$H(\omega) = (1 - d) + d e^{i\omega \Delta t}. \tag{5.28}$$

Let's not worry about a phase shift here; so we look at $|H(\omega)|^2$:

$$
\begin{aligned}
|H(\omega)|^2 &= H(\omega) H^*(\omega) \\
&= (1 - d + d \cos \omega \Delta t + i d \sin \omega \Delta t)(1 - d + d \cos \omega \Delta t - i d \sin \omega \Delta t) \\
&= (1 - d + d \cos \omega \Delta t)^2 + d^2 \sin^2 \omega \Delta t \\
&= 1 - 2d + d^2 + d^2 \cos^2 \omega \Delta t + 2(1 - d)d \cos \omega \Delta t + d^2 \sin^2 \omega \Delta t \\
&= 1 - 2d + 2d^2 + 2(1 - d)d \cos \omega \Delta t \\
&= 1 + 2d(d - 1) - 2d(d - 1) \cos \omega \Delta t \\
&= 1 + 2d(d - 1)(1 - \cos \omega \Delta t) \\
&= 1 + 4d(d - 1) \sin^2 \frac{\omega \Delta t}{2} \\
&= 1 - 4d(1 - d) \sin^2 \frac{\omega \Delta t}{2}.
\end{aligned}
\tag{5.29}
$$

The function $|H(\omega)|^2$ is shown in Fig. 5.9 for $d = 0$, $d = 1/4$ and $d = 1/2$.

This means: apart from the (not unexpected) phase shift, we have a low-pass effect due to the interpolation, similar to what happened in (5.12), which is strongest for $d = 1/2$. If we know that our sampled function $f(t)$ is bandwidth-limited, we may

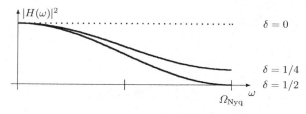

Fig. 5.9 Modulus squared of the transfer function for the shifting-/interpolation-algorithm (5.27)

use the sampling theorem and perform the "correct" interpolation, without getting a low-pass effect. Reconstructing $f(t)$ from samples f_k, however, requires quite an effort and often is not necessary. Interpolation algorithms requiring much effort are either not necessary (in case the relevant spectral components are markedly below Ω_{Nyq}), or they easily result in high-frequency artefacts. So be careful! Boundary effects have to be treated separately.

5.4 Data Compression

Often we get the problem where data sampling had been too fine, so data have to be compressed. An obvious algorithm would be, for example:

$$y_j \equiv y_{2k} = \frac{1}{2}(f_k + f_{k+1}), \ j = 0, \ldots, \frac{N}{2} \quad \text{"compression"}. \tag{5.30}$$

Here data set $\{y_k\}$ is only half as long as data set $\{f_k\}$. We pretend to have extended the sampling width Δt by the factor 2 and expect the average of the old samples at the sampling point. This inevitably will lead to a phase shift:

$$H(\omega) = \frac{1}{2} + \frac{1}{2}e^{i\Delta t}. \tag{5.31}$$

If we don't want that, we better use the smoothing-algorithm (5.12), where only every other output is stored:

$$y_j \equiv y_{2k} = \frac{1}{4}(f_{k-1} + 2f_k + f_{k+1}), \ j = 0, \ldots, \frac{N}{2} \quad \text{"compression"}. \tag{5.32}$$

Here, there is no phase shift, the principle is shown in Fig. 5.10.
Boundary effects have to be treated separately.
So we might assume, for example, $f_{-1} = f_0$ for the calculation of y_0. This also applies to the end of the data set.

Fig. 5.10 Data compression algorithm of (5.32)

5.5 Differentiation of Discrete Data

We may define the derivative of a sampled function as:

$$\frac{\mathrm{d}f}{\mathrm{d}t} \equiv y_k = \frac{f_{k+1} - f_k}{\Delta t} \quad \text{"first forward difference"}. \tag{5.33}$$

The corresponding transfer function reads:

$$\begin{aligned}
H(\omega) &= \tfrac{1}{\Delta t}\left(e^{i\omega\Delta t} - 1\right) = \tfrac{1}{\Delta t}e^{i\omega\Delta t/2}\left(e^{i\omega\Delta t/2} - e^{-i\omega\Delta t/2}\right)\\
&= \tfrac{2i}{\Delta t}\sin\tfrac{\omega\Delta t}{2}e^{i\omega\Delta t/2}\\
&= i\omega e^{i\omega\Delta t/2}\tfrac{\sin\frac{\omega\Delta t}{2}}{\omega\Delta t/2}.
\end{aligned} \tag{5.34}$$

The exact result would be $H(\omega) = i\omega$ (cf. (2.56)), the second and the third factor are due to the discretisation. The phase shift in (5.34) is a nuisance.

The "first backward difference":

$$y_k = \frac{f_k - f_{k-1}}{\Delta t}. \tag{5.35}$$

has got the same problem. The "first central difference":

$$y_k = \frac{f_{k+1} - f_{k-1}}{2\Delta t} \tag{5.36}$$

solves the problem with the phase shift. Here the following applies:

$$\begin{aligned}
H(\omega) &= \frac{1}{2\Delta t}\left(e^{+i\omega\Delta t} - e^{-i\omega\Delta t}\right)\\
&= i\omega\frac{\sin\omega\Delta t}{\omega\Delta t}.
\end{aligned} \tag{5.37}$$

Here, however, the filter effect is more pronounced, as is shown in Fig. 5.11.

For high frequencies the derivative becomes more and more wrong.

Fix: Sample as fine as possible, so that within your frequency realm $\omega \ll \Omega_{\mathrm{Nyq}}$ is always true.

Fig. 5.11 Transfer function of the "first central difference" (5.36) and the exact value (*thin line*)

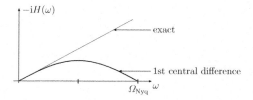

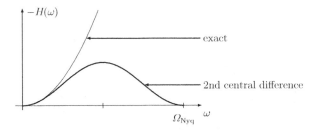

Fig. 5.12 Transfer function of the "second central difference" (5.39) and exact value (*thin line*)

The "second central difference" is as follows:

$$y_k = \frac{f_{k-2} - 2f_k + f_{k+2}}{4\Delta t^2}.$$

(5.38)

It corresponds to the second derivative. The corresponding transfer function is as follows:

$$
\begin{aligned}
H(\omega) &= \frac{1}{4\Delta t^2}\left(e^{-i\omega 2\Delta t} - 2 + e^{+i\omega 2\Delta t}\right) \\
&= \frac{1}{4\Delta t^2}(2\cos 2\omega\Delta t - 2) = -\frac{1}{\Delta t^2}\sin^2\omega\Delta t \\
&= -\omega^2\left(\frac{\sin\omega\Delta t}{\omega\Delta t}\right)^2.
\end{aligned}
$$

(5.39)

This should be compared to the exact expression $H(\omega) = (i\omega)^2 = -\omega^2$. Figure 5.12 shows $-H(\omega)$ for both cases.

5.6 Integration of Discrete Data

The simplest way to "integrate" data is to sum them up. It's a bit more precise if we interpolate between the data points. Let's use the Trapezoidal Rule as an example: assume the sum up to the index k to be y_k, in the next step we add the following trapezoidal area (cf. Fig. 5.13):

$$y_{k+1} = y_k + \frac{\Delta t}{2}(f_{k+1} + f_k) \qquad \text{"Trapezoidal Rule".}$$

(5.40)

The algorithm is: $(V^1 - 1)\, y_k = (\Delta t/2)\,(V^1 + 1)\, f_k$, V^l is the shifting operator of (5.4).

Fig. 5.13 Concerning the
trapezoidal rule

So the corresponding transfer function is:

$$
\begin{aligned}
H(\omega) &= \frac{\Delta t}{2} \frac{\left(e^{i\omega\Delta t} + 1\right)}{\left(e^{i\omega\Delta t} - 1\right)} \\
&= \frac{\Delta t}{2} \frac{e^{i\omega\Delta t/2}\left(e^{+i\omega\Delta t/2} + e^{-i\omega\Delta t/2}\right)}{e^{i\omega\Delta t/2}\left(e^{+i\omega\Delta t/2} - e^{-i\omega\Delta t/2}\right)} \\
&= \frac{\Delta t}{2} \frac{2\cos(\omega\Delta t/2)}{2i\sin(\omega\Delta t/2)} = \frac{1}{i\omega}\frac{\omega\Delta t}{2}\cot\frac{\omega\Delta t}{2}.
\end{aligned}
\tag{5.41}
$$

The "exact" transfer function is:

$$
H(\omega) = \frac{1}{i\omega} \qquad \text{see also (2.63)}.
\tag{5.42}
$$

Heaviside's step function has the Fourier transform $1/i\omega$, we get that when integrating over the impulse (δ-function) as input. The factor $(\omega\Delta t/2)\cot(\omega\Delta t/2)$ is due to the discretization. $H(\omega)$ is shown in Fig. 5.14.

The Trapezoidal Rule is a very useful integration algorithm.

Another integration algorithm is Simpson's 1/3-rule, which can be derived as follows.

Given are three subsequent numbers f_0, f_1, f_2 and we want to put a second order polynomial through these points:

$$
\begin{aligned}
y &= a + bx + cx^2 \\
\text{with } y(x = 0) &= f_0 = a, \\
y(x = 1) &= f_1 = a + b + c, \\
y(x = 2) &= f_2 = a + 2b + 4c .
\end{aligned}
\tag{5.43}
$$

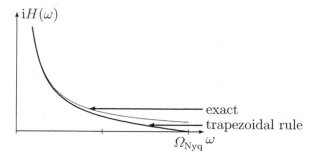

Fig. 5.14 Transfer function for the trapezoidal rule (5.40) and exact value (*thin line*)

The resulting coefficients are:

$$
\begin{aligned}
a &= f_0, \\
c &= f_0/2 + f_2/2 - f_1, \\
b &= f_1 - f_0 - c = f_1 - f_0 - f_0/2 - f_2/2 + f_1 \\
&= 2f_1 - 3f_0/2 - f_2/2.
\end{aligned}
\tag{5.44}
$$

The integration of this polynomial of $0 \le x \le 2$ results in:

$$
\begin{aligned}
I &= 2a + 4\frac{b}{2} + 8\frac{c}{3} \\
&= 2f_0 + 4f_1 - 3f_0 - f_2 + \frac{4}{3}f_0 + \frac{4}{3}f_2 - \frac{8}{3}f_1 \\
&= \frac{1}{3}f_0 + \frac{4}{3}f_1 + \frac{1}{3}f_2 = \frac{1}{3}(f_0 + 4f_1 + f_2).
\end{aligned}
\tag{5.45}
$$

This is called Simpson's 1/3-rule. As we've gathered up $2\Delta t$, we need the step-width $2\Delta t$. So the algorithm is:

$$
y_{k+2} = y_k + \frac{\Delta t}{3}(f_{k+2} + 4f_{k+1} + f_k) \qquad \text{"Simpson's 1/3-rule"}.
\tag{5.46}
$$

This corresponds to an interpolation with a second order polynomial. The transfer function is:

$$
H(\omega) = \frac{1}{i\omega}\frac{\omega\Delta t}{3}\frac{2 + \cos\omega\Delta t}{\sin\omega\Delta t}
$$

and is shown in Fig. 5.15.

At high frequencies Simpson's 1/3-rule gives grossly wrong results. Of course, Simpson's 1/3-rule is more exact than the Trapezoidal Rule, given medium frequencies, or the effort of interpolation with a second order polynomial would be hardly worth it.

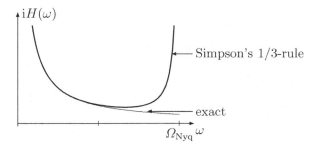

Fig. 5.15 Transfer function for Simpson's 1/3-rule compared to the Trapezoidal Rule and the exact value (*thin line*)

At $\omega = \Omega_{\mathrm{Nyq}}/2$ we have, relative to $H(\omega) = 1/i\omega$:

Trapezoid:

$$\frac{\Omega_{\mathrm{Nyq}}\Delta t}{4} \cot \frac{\Omega_{\mathrm{Nyq}}\Delta t}{4} = \frac{\pi}{4} \cot \frac{\pi}{4} = \frac{\pi}{4} = 0.785 \quad (\text{too small}),$$

Simpson's-1/3:

$$\frac{\Omega_{\mathrm{Nyq}}\Delta t}{6} \frac{2 + \cos(\Omega_{\mathrm{Nyq}}\Delta t/2)}{\sin(\Omega_{\mathrm{Nyq}}\Delta t/2)} = \frac{\pi}{6} \frac{2+0}{1} = \frac{\pi}{3} = 1.047 \quad (\text{too big}).$$

Simpson's 1/3-rule also does better for low frequencies than the Trapezoidal Rule:

Trapezoid:

$$\frac{\omega \Delta t}{2} \left(\frac{1}{\omega \Delta t/2} - \frac{\omega \Delta t/2}{3} + \dots \right) \approx 1 - \frac{\omega^2 \Delta t^2}{12},$$

Simpson's-1/3:

$$\frac{\omega \Delta t}{3} \frac{\left(2 + 1 - \dfrac{1}{2}\omega^2 \Delta t^2 + \dfrac{\omega^4 \Delta t^4}{24} \cdots \right)}{\omega \Delta t \left(1 - \dfrac{\omega^2 \Delta t^2}{6} + \dfrac{\omega^4 \Delta t^4}{120} \cdots \right)}$$

$$= \frac{1 - \dfrac{\omega^2 t^2}{6} + \dfrac{\omega^4 t^4}{72} \cdots}{1 - \dfrac{\omega^2 t^2}{6} + \dfrac{\omega^4 t^4}{120} \cdots} \approx 1 + \frac{\omega^4 \Delta t^4}{180} \cdots .$$

The examples in Sects. 5.2–5.6 would point us in the following direction, as far as digital data processing is concerned:

The rule of thumb therefore is:

Do sample as fine as possible!
Keep away from Ω_{Nyq}!

Do also try out other algorithms, and have lots of fun!

Playground

5.1 Totally Different
Given is the function $f(t) = \cos(\pi t/2)$ which is sampled at times $t_k = k\Delta t$, $k = 0, 1, \ldots, 5$ with $\Delta t = 1/3$.

Calculate the first central difference and compare it with the "exact" result for $f'(t)$. Plot your results! What is the percentage error?

5.2 Simpson's-1/3 versus Trapezoid
Given is the function $f(t) = \cos \pi t$ which is sampled at times $t_k = k\Delta t$, $k = 0, 1, \ldots, 4$ with $\Delta t = 1/3$.

Calculate the integral using the Simpson's 1/3-rule and the Trapezoidal Rule and compare your results with the exact value.

5.3 Totally Noisy
Given is a cosine input series that's practically smothered by noise (Fig. 5.16).

$$f_i = \cos \frac{\pi j}{4} + 5(\text{RND} - 0.5), \qquad j = 0, 1, \ldots, N. \qquad (5.47)$$

In our example, the noise has a 2.5-times higher amplitude than the cosine signal. (The signal-to-noise ratio (power!) therefore is $0.5 : 5/12 = 1.2$, see playground 4.6.)

In the time spectrum (Fig. 5.16) we can't even guess the existence of the cosine component.

Fig. 5.16 Cosine signal in totally noisy background according to (5.47)

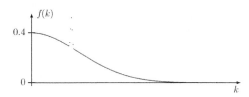

Fig. 5.17 Discrete line on slowly falling background

a. What Fourier transform do you expect for series (5.47)?

b. What can you do to make the cosine component visible in the time spectrum, too?

5.4 Inclined Slope

Given is a discrete line as input, that's sitting on a slowly falling ground (Fig. 5.17).

a. What's the most elegant way of getting rid of the background?

b. How do you get rid of the "undershoot"?

Chapter 6
Data Streams and Fractional Delays

Abstract In this chapter the concept of data streams produced by fast digitizers is introduced. Two simple filters to accomplish fractional delays are presented: the Lagrange interpolator, a non-recursive filter, and the Thiran all-pass filter, a recursive filter, both in lowest order. The group delay of these filters is discussed in detail and illustrated with examples. For the Thiran all-pass the impulse, step, and ramp response are shown.

This and the following chapter are for those pedestrians who want to go a little further, like the German writer Johann Gottfried Seume who *walked* from Grimma, a little town near Leipzig, Free State of Saxony, Germany, to Syrakus in Sicily, Siracusa in Italian ("Spaziergang nach Syrakus").

Thus far we have considered a static set of data on which we carry out actions like high- and low-pass filtering or alike. Modern digitizers analyze an analogue signal at a certain pace, e.g. every nanosecond, and provide a stream of digital output. Light travels about 30 cm in a nanosecond, the typical length of a ruler. Quite a handy distance, indeed! Such digitizers are in use in a variety of fields, e.g. in acoustics or microwave antenna arrays. Since there are different propagation delays involved in cables for loudspeakers or antennas, the need for delays is immediately apparent. Integer delays pose no problem: a simple shift register, a so-called FIFO (first in, first out) would provide the solution. However, what about fractional delays?

6.1 Fractional Delays

We could simply interpolate the data like we did in Sect. 5.3 called "shifting data". This is indeed a possibility and we should examine it in more detail. The linear interpolation is merely the simplest version of the so-called Lagrange-interpolation schemes which use polynomials for interpolation. An example for a quadratic interpolation ($N = 2$) is Simpson's 1/3-rule.

© Springer International Publishing Switzerland 2015

T. Butz, *Fourier Transformation for Pedestrians*, Undergraduate Lecture
Notes in Physics, DOI 10.1007/978-3-319-16985-9_6

6.2 Non-recursive Algorithms

Non-recursive algorithms interpolate past and present input data $f_{-N}, f_{-N+1}, \ldots, f_0$ by polynomials of degree N. The simplest non-recursive algorithm for a fractional delay is the Lagrange interpolator for $N = 1$, i.e. a linear interpolation. For this case we noticed that the amplitude of an input signal is reduced depending on the delay and the frequency (see Fig. 5.9). Thus, e.g. the Nyquist frequency is not transmitted at all for $d = 0.5$. Since we are dealing with data streams we call the pace Δt the "tag" and interpret d as a delay. Moreover, we did not pay attention to the associated phase shift.

Here, we shall first consider the phase shift, but in a rather different way. Instead of speaking of the phase ϕ we shall use the quantity "group delay" defined as:

$$\tau_{\text{group}} = -\frac{d\phi}{d\omega}. \tag{6.1}$$

This is the analogue of the inverse group velocity—e.g. in light guides—per unit length. It measures the time delay of the amplitude envelope of a signal consisting of various sinusoidal components produced by a linear algorithm with time invariance.

Let us have a look at the Lagrange interpolator with $N = 1$. The difference equation is written as:

$$y_k = df_{k-1} + (1 - d)f_k. \tag{6.2}$$

Note that we have slightly changed the nomenclature compared to (5.27) because we now have to use terms like "past" and "present" and we do not want to use "future" in a data stream (although this would be possible, of course, by using a certain overall integer delay to convert "future" into "past"). The range of d is simply $0 \le d \le 1$. The magnitude of the transfer function $H(\omega)$ is not affected by this change in nomenclature.

Let us see what we need:
First, we write the transfer function corresponding to (5.28) in the following way:

$$H(z) = dz^{-1} + (1 - d) \tag{6.3}$$

with $z = e^{i\omega \Delta t}$, a convenient abbreviation for the moment. We shall come back to this point later. This equation just means that you have to take the past quantity multiplied by d and the present quantity multiplied by $(1 - d)$. In this sense, z^{-1} has the same function as our shift operator V^{-1} defined in (5.20). The real and imaginary parts of $H(z)$ are:

$$\begin{aligned} \text{Re}\{H(z)\} &= d\cos\omega\Delta t + (1 - d) \\ \text{Im}\{H(z)\} &= -d\sin\omega\Delta t. \end{aligned} \tag{6.4}$$

Hence, the phase is:

$$\phi = \arctan \frac{-d \sin \omega \Delta t}{d \cos \omega \Delta t + (1-d)}. \tag{6.5}$$

Differentiating the arctan-function with respect to ω is a little tedious. With the help of:

$$\frac{d \arctan \alpha}{d\alpha} = \frac{1}{1+\alpha^2} \tag{6.6}$$

and a few simplifications we arrive at:

$$\tau_{\text{group}} = \frac{d\Delta t \left(1 - 2(1-d)\sin^2 \frac{\omega \Delta t}{2}\right)}{\left(1 - 4d(1-d)\sin^2 \frac{\omega \Delta t}{2}\right)}. \tag{6.7}$$

This function is plotted in Fig. 6.1 versus ω for various values of d with $\Delta t = 1$.

The limiting values for $\omega \to 0$ are easily seen:

$$\tau_{\text{group}} \to d \qquad \text{(in units of } \Delta t\text{)}. \tag{6.8}$$

For $\omega \to \Omega_{\text{Nyq}}$ we get:

$$\tau_{\text{group}} \to \frac{d}{2d-1} \qquad \text{(in units of } \Delta t\text{)}. \tag{6.9}$$

For $d = 1/2$ the group delay seems to diverge at Ω_{Nyq}. Had we first inserted $d = 1/2$ into (6.7), the frequency dependent terms in the numerator and denominator would cancel and we would get $\tau_{\text{group}} = 1/2$ in the entire frequency range. Thus, the result depends on the order of taking the limiting values. What a nuisance! It does not really help that under these conditions $H(\Omega_{\text{Nyq}}) = 0$, i.e. nothing is passed at all. The problem is still severe in the vicinity of Ω_{Nyq} for d close to $1/2$. What is even more intriguing is that τ_{group} becomes negative for $d < 1/2$ at a sufficiently high frequency (always above $\Omega_{\text{Nyq}}/2$). What is the meaning of a "negative delay"? Don't

Fig. 6.1 The group delay τ_{group} in units of Δt for the $N = 1$ Lagrange interpolator versus ω for delays d in the range from 0 to 1

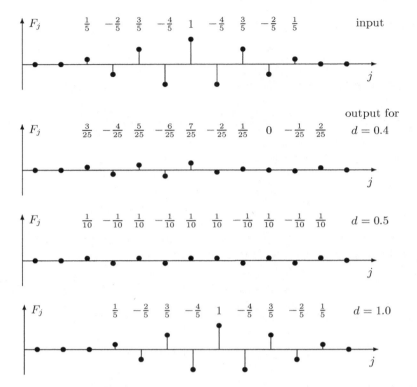

Fig. 6.2 Input signal with $\omega = \Omega_{\text{Nyq}}$ and triangular envelope (*top*); output of the Lagrange interpolator with $N = 1$ for various delays d (subsequent to *top*)

panic! For the envelope of a signal which varies slowly compared to the frequency components there is no problem in "advancing" the maximum a few tags like it is for delaying. In order to see what cruelties the seemingly "innocent" Lagrange interpolater with $N = 1$ can commit, let's do a "drastic" example.

Example 6.1 ("Nyquist pulse"). We use as input a signal with $\omega = \Omega_{\text{Nyq}}$ and a triangular envelope (see Fig. 6.2(*top*)). The signal contains spectral components like those shown in Fig. 3.2, now centered at Ω_{Nyq}. The upper half is simply the mirrored left half like illustrated in Fig. 4.5. This choice makes sure that we cover the range of frequencies where τ_{group} is strongly frequency dependent, except for $d = 0.5$. Figure 6.2 illustrates the output for the delays $d = 0.4, 0.5$, and 1. For $d = 0.5$ we immediately see that we get a delay of 0.5, the pulse envelope is now rectangular and the amplitude is heavily attenuated. Where is the maximum of the envelope? In the middle? Since $\tau_{\text{group}} = 0.5$, independent of ω, the rectangular envelope is entirely the result of the frequency dependent attenuation. For $d = 0.4$ the signal is heavily distorted and the envelope is monotonous on one side only. It looks like the envelope is shifted backwards. For $d = 0.6$ we get the mirror image of the case for $d = 0.4$. For $d = 1$ we recover the input shifted by 1.

We see that not only the magnitude of the transfer function $H(\omega)$ depends on d and ω but also the group delay. A rather unsatisfactory situation.

Thankfully, there is help regarding the magnitude of the transfer function using a recursive algorithm, as we shall see in Sect. 6.4.

In order to see that the Lagrange interpolator for $N = 1$ is not dull at all, let's do two examples.

Example 6.2 (Average) Its worth looking at a subtle detail: whats the average of the function with the Nyquist frequency and a triangular envelope (see Fig. 6.2(*top*))? Have a guess! Looks like zero. Not bad, but not good enough! It is zero for even M only where M is the number of data points below 0 and also above 0, i.e. there are $2M$ data points. What is it for $M = $ odd ? Lets do the calculation by "brute force":

$$
\begin{aligned}
\text{Average} &= \frac{1}{2M}\left(1 + 2\sum_{k=1}^{M-1}(-1)^k\frac{M-k}{M}\right) \\
&= \frac{1}{2M}\left(1 + 2\sum_{k=1}^{M-1}(-1)^k - \frac{2}{M}\sum_{k=1}^{M-1}(-1)^k k\right) \\
&= \frac{1}{2M}\left(1 + \left\{\begin{matrix}-2 \\ 0\end{matrix}\right\} - \frac{2}{M}\left\{\begin{matrix}\frac{M-2}{2} - (M-1) \\ \frac{M-1}{2}\end{matrix}\right\}\right) \begin{matrix}\text{for even } M \\ \text{for odd } M\end{matrix} \\
&= \frac{1}{2M}\left(1 + \left\{\begin{matrix}-2 \\ 0\end{matrix}\right\} - \frac{2}{M}\left\{\begin{matrix}-\frac{M}{2} \\ \frac{M-1}{2}\end{matrix}\right\}\right) \begin{matrix}\text{for even } M \\ \text{for odd } M\end{matrix} \\
&= \frac{1}{2M}\left(1 + \left\{\begin{matrix}-2 \\ 0\end{matrix}\right\} + \left\{\begin{matrix}1 \\ \frac{1}{M}-1\end{matrix}\right\}\right) \begin{matrix}\text{for even } M \\ \text{for odd } M\end{matrix} \\
&= \left\{\begin{matrix}0 & \text{for even } M \\ \frac{1}{2M^2} & \text{for odd } M\end{matrix}\right. .
\end{aligned}
$$

It has to do with the question whether the first/last non-vanishing data point is negative or positive. Of course, for large M the difference fades away. A look to Fourier space will show that this result comes in quite naturally. To simplify matters, we use periodic continuation, i.e. no zero-padding. Lets start with the calculation of the discrete Fourier transform of the discrete triangular function. I'm sure you have noticed that in Playground 4.9 we have already solved a similar problem. The triangular function— or double-sided ramp—was shifted and we required even N. With a slight change in nomenclature we can relax this requirement. Should you happen to have solved Playground 4.9 take the following as a rehearsal. Let's start from scratch.

$$f_k = \begin{cases} \dfrac{M+k}{M} & \text{for } -M \le k \le 0 \\[2mm] \dfrac{M-k}{M} & \text{for } 1 \le k \le M-1 \end{cases}$$

$$F_0 = \frac{1}{2M}\left(1 + 2\sum_{k=1}^{M-1}\left(1 - \frac{k}{M}\right)\right)$$

$$= \frac{1}{2M}\left(1 + 2(M-1) - \frac{2}{M}\sum_{k=1}^{M-1}k\right)$$

$$= \frac{1}{2M}\left(1 + 2M - 2 - \frac{2}{M}\frac{M(M-1)}{2}\right)$$

$$= \frac{1}{2M}(1 + 2M - 2 - M + 1) = \frac{1}{2},$$

as we would have guessed right away.

The second sum is due to the young mathematician Carl Friedrich Gauß who apparently did not want to sum up all numbers from 1 to 100 step by step, a task given by the teacher to keep the pupils busy. Carl Friedrich noticed that by complementing each number to add up to the maximum number he actually adds the same series but in reversed order. That's where the factor of two in the denominator comes from.

For $j > 0$ we have:

$$F_j = \frac{1}{2M}\left(\sum_{k=-M}^{-1} f_k \exp\frac{-i\pi jk}{M} + 1 + \sum_{k=1}^{M-1} f_k \exp\frac{-i\pi jk}{M}\right)$$

$$= \frac{1}{2M}\left(1 + 2\sum_{k=1}^{M-1}\left(\left(1 - \frac{k}{M}\right)\cos\frac{\pi jk}{M}\right)\right)$$

(term with $k = -M$ vanishes because $f_{-M} = 0$)

$$= \frac{1}{2M}\left(1 + 2\sum_{k=1}^{M-1}\cos\frac{\pi jk}{M} - \frac{2}{M}\sum_{k=1}^{M-1}k\cos\frac{\pi jk}{M}\right).$$

The first sum is easily evaluated using the Dirichlet kernel (1.53) with the upper summation index being $M - 1$. We get:

$$2\sum_{k=1}^{M-1}\cos\frac{\pi jk}{M} = 2\left(\frac{\sin\left(M - \frac{1}{2}\right)x}{2\sin\frac{x}{2}} - \frac{1}{2}\right)$$

$$= -\cos Mx - 1 \quad \text{with } x = \frac{\pi j}{M}.$$

For the second sum we start with the expression derived in Playground 4.9:

$$\sum_{k=1}^{M-1}\sin kx = \frac{\cos\frac{x}{2} - \cos\left(M - \frac{1}{2}\right)x}{2\sin\frac{x}{2}} = \frac{1}{2}\left((1 - \cos Mx)\cot\frac{x}{2} - \sin Mx\right)$$

is shifted backwards. For $d = 0.6$ we get the mirror image of the case for $d = 0.4$. For $d = 1$ we recover the input shifted by 1.

We see that not only the magnitude of the transfer function $H(\omega)$ depends on d and ω but also the group delay. A rather unsatisfactory situation.

Thankfully, there is help regarding the magnitude of the transfer function using a recursive algorithm, as we shall see in Sect. 6.4.

In order to see that the Lagrange interpolator for $N = 1$ is not dull at all, let's do two examples.

Example 6.2 (Average) Its worth looking at a subtle detail: whats the average of the function with the Nyquist frequency and a triangular envelope (see Fig. 6.2(*top*))? Have a guess! Looks like zero. Not bad, but not good enough! It is zero for even M only where M is the number of data points below 0 and also above 0, i.e. there are $2M$ data points. What is it for $M = $ odd ? Lets do the calculation by "brute force":

$$
\begin{aligned}
\text{Average} &= \frac{1}{2M}\left(1 + 2\sum_{k=1}^{M-1}(-1)^k \frac{M-k}{M}\right)\\[2mm]
&= \frac{1}{2M}\left(1 + 2\sum_{k=1}^{M-1}(-1)^k - \frac{2}{M}\sum_{k=1}^{M-1}(-1)^k k\right)\\[2mm]
&= \frac{1}{2M}\left(1 + \begin{Bmatrix}-2\\0\end{Bmatrix} - \frac{2}{M}\begin{Bmatrix}\frac{M-2}{2}-(M-1)\\\frac{M-1}{2}\end{Bmatrix}\right)\begin{matrix}\text{for even }M\\\text{for odd }M\end{matrix}\\[2mm]
&= \frac{1}{2M}\left(1 + \begin{Bmatrix}-2\\0\end{Bmatrix} - \frac{2}{M}\begin{Bmatrix}-\frac{M}{2}\\\frac{M-1}{2}\end{Bmatrix}\right)\begin{matrix}\text{for even }M\\\text{for odd }M\end{matrix}\\[2mm]
&= \frac{1}{2M}\left(1 + \begin{Bmatrix}-2\\0\end{Bmatrix} + \begin{Bmatrix}1\\\frac{1}{M}-1\end{Bmatrix}\right)\begin{matrix}\text{for even }M\\\text{for odd }M\end{matrix}\\[2mm]
&= \begin{cases}0 & \text{for even }M\\\frac{1}{2M^2} & \text{for odd }M\end{cases}.
\end{aligned}
$$

It has to do with the question whether the first/last non-vanishing data point is negative or positive. Of course, for large M the difference fades away. A look to Fourier space will show that this result comes in quite naturally. To simplify matters, we use periodic continuation, i.e. no zero-padding. Lets start with the calculation of the discrete Fourier transform of the discrete triangular function. I'm sure you have noticed that in Playground 4.9 we have already solved a similar problem. The triangular function— or double-sided ramp—was shifted and we required even N. With a slight change in nomenclature we can relax this requirement. Should you happen to have solved Playground 4.9 take the following as a rehearsal. Let's start from scratch.

$$f_k = \begin{cases} \dfrac{M+k}{M} & \text{for } -M \leq k \leq 0 \\[2mm] \dfrac{M-k}{M} & \text{for } 1 \leq k \leq M-1 \end{cases}$$

$$F_0 = \frac{1}{2M}\left(1 + 2\sum_{k=1}^{M-1}\left(1 - \frac{k}{M}\right)\right)$$

$$= \frac{1}{2M}\left(1 + 2(M-1) - \frac{2}{M}\sum_{k=1}^{M-1}k\right)$$

$$= \frac{1}{2M}\left(1 + 2M - 2 - \frac{2}{M}\frac{M(M-1)}{2}\right)$$

$$= \frac{1}{2M}(1 + 2M - 2 - M + 1) = \frac{1}{2},$$

as we would have guessed right away.

The second sum is due to the young mathematician Carl Friedrich Gauß who apparently did not want to sum up all numbers from 1 to 100 step by step, a task given by the teacher to keep the pupils busy. Carl Friedrich noticed that by complementing each number to add up to the maximum number he actually adds the same series but in reversed order. That's where the factor of two in the denominator comes from.

For $j > 0$ we have:

$$F_j = \frac{1}{2M}\left(\sum_{k=-M}^{-1} f_k \exp\frac{-i\pi jk}{M} + 1 + \sum_{k=1}^{M-1} f_k \exp\frac{-i\pi jk}{M}\right)$$

$$= \frac{1}{2M}\left(1 + 2\sum_{k=1}^{M-1}\left(\left(1 - \frac{k}{M}\right)\cos\frac{\pi jk}{M}\right)\right)$$

(term with $k = -M$ vanishes because $f_{-M} = 0$)

$$= \frac{1}{2M}\left(1 + 2\sum_{k=1}^{M-1}\cos\frac{\pi jk}{M} - \frac{2}{M}\sum_{k=1}^{M-1}k\cos\frac{\pi jk}{M}\right).$$

The first sum is easily evaluated using the Dirichlet kernel (1.53) with the upper summation index being $M - 1$. We get:

$$2\sum_{k=1}^{M-1}\cos\frac{\pi jk}{M} = 2\left(\frac{\sin\left(M - \frac{1}{2}\right)x}{2\sin\frac{x}{2}} - \frac{1}{2}\right)$$

$$= -\cos Mx - 1 \quad \text{with } x = \frac{\pi j}{M}.$$

For the second sum we start with the expression derived in Playground 4.9:

$$\sum_{k=1}^{M-1}\sin kx = \frac{\cos\frac{x}{2} - \cos\left(M - \frac{1}{2}\right)x}{2\sin\frac{x}{2}} = \frac{1}{2}\left((1 - \cos Mx)\cot\frac{x}{2} - \sin Mx\right)$$

This we differentiate with respect to x to get:

$$\sum_{k=1}^{M-1} k \cos kx = \frac{1}{2} \left(\frac{-(1 - \cos Mx)\frac{1}{2}}{\sin^2 \frac{x}{2}} + M \sin Mx \cot \frac{x}{2} - M \cos Mx \right)$$

$$= -\frac{1}{2} \frac{M \cos \pi j + \sin^2 \frac{\pi j}{2}}{\sin^2 \frac{\pi j}{2M}} \quad \text{with } x = \frac{\pi j}{M}.$$

Collecting all terms we finally arrive at:

$$F_j = \frac{1}{2M^2} \frac{\sin^2 \frac{\pi j}{2}}{\sin^2 \frac{\pi j}{2M}}.$$

This is the discrete analogue of the continuous result of (3.4). We can simplify this expression further to:

$$F_0 = \frac{1}{2} \quad \text{(the average;}$$

we can use even the general formula,
let j be continuous for the moment,
and use l'Hospital's rule)

$$F_j = \begin{cases} \dfrac{1}{2M^2} \dfrac{1}{\sin^2 \frac{\pi j}{2M}} & \text{for odd } j \\ 0 & \text{for even } j \end{cases}.$$

The continuous Fourier transform of the triangular function had zeros so does the discrete one. What is left over is to shift the Fourier transform of the discrete triangular function to the Nyquist frequency because a multiplication of the function with the Nyquist frequency by the triangular function corresponds to the convolution of the Fourier-transformed triangular function with a discrete δ-function at the Nyquist frequency. For the moment we are only interested in the value at zero frequency because this gives the average of our function where we started. This means we need the value at the end of the tail, i.e. $j = -M$ for the unshifted Fourier-transformed triangular function. We get:

$$F_0 = \begin{cases} \dfrac{1}{2M^2} \dfrac{\sin^2 \frac{\pi M}{2}}{\sin^2 \frac{\pi}{2}} = \dfrac{1}{2M^2} & \text{for odd } M \\ 0 & \text{for even } M \end{cases}.$$

Now it is clear that the average of our input function is non-zero for odd M only.

Example 6.3 (Frequency comb) Now lets have a closer look at the case $d = 1/2$ in Fourier space. The function looks like the Nyquist frequency, but there is a flaw in the middle. In fact, this little glitch is causing a whole comb of non-zero frequency components in Fourier space, as we'll see below. First we need the discrete transfer function for the Lagrange interpolator with $N = 1$. Since we are not interested in the phase shift, we need the real part only: $\text{Re}\{H_j\} = \cos^2(\pi j/2M)$. For the sake of convenience, we'll drop the $\text{Re}\{\}$ in the following. This we multiply by the Fourier-transformed triangular function centred at the Nyquist frequency. The shifted Fourier transform of the triangular function reads:

$$F_0 = \begin{cases} \dfrac{1}{2M^2} & \text{for odd } M \\[2mm] 0 & \text{for even } M \end{cases}$$

$$F_M = \frac{1}{2}$$

$$F_j = \begin{cases} \dfrac{1}{2M^2} \dfrac{1}{\sin^2 \frac{\pi(M-j)}{2M}} & \begin{array}{l} \text{for odd } M \text{ and even } j = 2, 4, \dots, M-1 \\ \text{or} \\ \text{for even } M \text{ and odd } j = 1, 3, \dots, M-1 \end{array} \\[4mm] 0 & \text{else} \end{cases}$$

$$F_{M+j} = F_{M-j} \quad \text{for } j = 1, 2, \dots, M-1.$$

The trigonometric expression can be further simplified:

$$\sin^2 \frac{\pi(M-j)}{2M} = \left(\overset{=1}{\sin\frac{\pi}{2}} \cos\frac{\pi j}{2M} - \overset{=0}{\cos\frac{\pi}{2}} \sin\frac{\pi j}{2M} \right)^2 = \cos^2 \frac{\pi j}{2M}.$$

No surprise, in the continuous world, $\sin^2 x$ shifted by $\pi/2$ yields $\cos^2 x$! However, there are a few subtleties in the digital world as you can see from the little extras F_0 and F_M. Incidentally (or not?), this is identical to the transfer function and thus the trigonometric expressions cancel. We finally get:

$$H_M F_M = \cos^2 \frac{\pi M}{2M} \frac{1}{2} = 0 \quad \text{(no transmission at the Nyquist frequency)}$$

$$H_j F_j = \begin{cases} \dfrac{1}{2M^2} & \text{for odd } M + j \\[2mm] 0 & \text{else} \end{cases}$$

$$H_{M+j} F_{M+j} = H_{M-j} F_{M-j} \quad \text{for } j = 0, \dots, M-1.$$

This expression means that every second Fourier coefficient vanishes, in particular the odd ones for odd M and the even ones for even M (see Fig. 6.3 for $M = 3$ and $M = 4$). And, amazingly, all non-zero coefficients have the same value $\frac{1}{2M^2}$. Thats

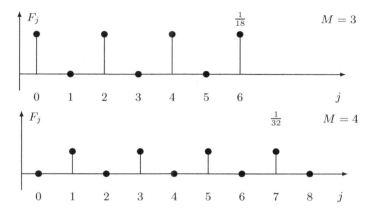

Fig. 6.3 Real part of Fourier transform of the input with the Nyquist frequency and triangular envelope delayed by $d = 1/2$ using the Lagrange interpolator with $N = 1$ for $M = 3$ (*top*) and $M = 4$ (*bottom*)

why we called this a "comb" of frequency coefficients. Its the glitch in the input data which caused it!

Just to complete the example, lets calculate directly the Fourier transform of the output of the function with the Nyquist frequency and the triangular envelope shifted by $d = 1/2$ using the Lagrange interpolator with $N = 1$. In order to simplify matters, we decompose the input into a constant with value $\frac{1}{2M}$ and a function with value $-\frac{1}{M}$ at those points where the input was negative. The constant is of no interest, it contributes to F_0 only and we know this value already ($=\frac{1}{2M^2}$ for odd M). The non-zero points of the second function appear at all odd negative values of k whereas they appear at all even k for positive ones. Since we are interested only in the real part of the Fourier transform we can flip around the negative k's to positive k's and see that we end up with a sum we know already: the Dirichlet kernel. There is a little pitfall: for even M we have to include the term with $k = -M$ whereas we must not for odd M. Lets discuss even and odd M separately. For even M we get:

$$
\begin{aligned}
F_j &= \frac{1}{2M}\left(-\frac{1}{M}\right)\left(\cos\pi j + \frac{\sin\left(M-\frac{1}{2}\right)\frac{\pi}{M}}{2\sin\frac{\pi j}{2M}} - \frac{1}{2}\right) \\
&= -\frac{1}{2M^2}\left(\cos\pi j - \frac{1}{2}\cos\pi j - \frac{1}{2}\right) \\
&= \frac{1}{2M^2}\sin^2\frac{\pi j}{2} \\
&= \begin{cases} \dfrac{1}{2M^2} & \text{for odd } j \\[2mm] 0 & \text{else} \end{cases} .
\end{aligned}
$$

For odd M we get:

$$F_j = \frac{1}{2M}\left(-\frac{1}{M}\right)\left(\frac{\sin\left(M-\frac{1}{2}\right)\frac{\pi}{M}}{2\sin\frac{\pi j}{2M}} - \frac{1}{2}\right)$$

$$= -\frac{1}{2M^2}\left(\frac{1}{2}\cos\pi j - \frac{1}{2}\right)$$

$$= \frac{1}{2M^2}\cos^2\frac{\pi j}{2}$$

$$= \begin{cases} \dfrac{1}{2M^2} & \text{for even } j \\ 0 & \text{else} \end{cases}.$$

This can be combined to:

$$F_j = \begin{cases} \dfrac{1}{2M^2} & \text{for odd } M+j \\ 0 & \text{else} \end{cases}$$

$$F_{M+j} = F_{M-j} \quad \text{for } j = 0,\ldots,M-1.$$

This is, of course, the same result as above.

6.3 Stability of Recursive Algorithms

In the Example 5.1 we have already encountered a recursive algorithm, also called a "filter". We can write the transfer function as a polynomial of the above defined quantity z^{-1} in the numerator and another polynomial in z^{-1} in the denominator, both with real-valued coefficients. We shall exclude positive powers of z because we want to use present, i.e. z^0, and past data only. For the discussion of stability of the filter we now allow z to be an arbitrary complex quantity rather than to be a phase only. Then we get N zeros of a polynomial of degree N in the denominator, hence the transfer function diverges. This is what is usually called a "pole". For $N = 1$ our denominator would read $1 + a_1 z^{-1}$ with a single pole for $z^{-1} = -1/a_1$ or $z = -a_1$. For higher N we would eventually get poles for complex values of z^{-1}. Let us examine what is the response of the transfer function:

$$H(z) = \frac{1}{1 - z^{-1}} \tag{6.10}$$

to an impulse:

$$f_k = \delta_{k,0}. \tag{6.11}$$

The corresponding difference equation reads:

$$y_k = y_{k-1}. \tag{6.12}$$

The output will be the Heaviside step function: once switched to unity it remains there forever. Sometimes (6.10) is called the "z-transform" of the Heaviside function. It is also called "accumulator" because it adds up all input data for $k \geq 0$.

If instead of (6.10) we would use:

$$H(z) = \frac{1}{1 - a_1 z^{-1}} \quad \text{with } |a_1| < 1 \tag{6.13}$$

we would have

$$y_k = a_1 y_{k-1} \tag{6.14}$$

and the response to the impulse would decay as a geometrical series with increasing powers of a_1. For small a_1, the output would decay rapidly, but actually persists indefinitely. This is why such filters are called "Infinite Impulse Response" (IIR) filters. The Lagrange interpolator, on the contrary, is a "Finite Impulse Response" (FIR) filter.

What has to be discussed now is the stability of the filter. We have seen that (6.10) is at the limit of stability while (6.13) is stable provided $|a_1| < 1$. It can be shown that IIR filters are stable provided the poles are all inside the unit circle in the complex plane. For the stability, the zeros of the numerator polynomial play no role.

6.4 Thiran's All-Pass Filter for $N = 1$

There is an algorithm which avoids the attenuation altogether, the so-called "Thiran all-pass filter", i.e. we have $|H(z)| = 1$ always, and, moreover, the group delay is "maximally flat", i.e. it depends only weakly on ω for small ω. More precisely, all up to the Nth derivatives at $\omega = 0$ are 0 for a polynomial of order N. Thiran [12] originally proposed a so-called "all-pole"-filter with a polynomial in z^{-1} in the denominator and a constant numerator. Fettweis [13] showed that it is advantageous to include a properly chosen polynomial in the numerator. This is now called the "Thiran all-pass filter" and is discussed in detail in [14].

You may ask, what the hell is filtered if everything passes? The "filter experts" call everything a "filter" which can be described by a linear difference equation, like the one above. It does not necessarily mean a filter in the frequency domain. You will get used to this terminology like you are used to click to the "start" button if you want to shut down your computer. I would have preferred a "stop" button.

Of course, we are dealing with a recursive filter.

Here is how it is constructed. We start with the transfer function:

$$H(z) = \frac{z^{-N} D_N(z)}{D_N(z^{-1})} \tag{6.15}$$

Here, $D_N(z)$ denotes a polynomial in z of order N with coefficients a_i and the constant $a_0 = 1$. It is easy to show that $|H(z)| = 1$.

The first factor z^{-N} is merely a phase with magnitude 1. The numerator can be written as $D_N(z) = |D_N(z)|e^{i\phi_D}$, the denominator can be written as $D_N(z^{-1}) = |D_N(z)|e^{-i\phi_D}$ because z^{-1} is its complex conjugate z^*. Thus the quotient of the two polynomials is again merely a phase $e^{2i\phi_D}$. This completes the proof.

Here, we shall discuss the simplest case with $N = 1$ only. We have:

$$H(z) = \frac{z^{-1}(1 + a_1 z)}{1 + a_1 z^{-1}} = \frac{z^{-1} + a_1}{1 + a_1 z^{-1}}. \tag{6.16}$$

The corresponding difference equation reads:

$$y_k = -a_1 y_{k-1} + f_{k-1} + a_1 f_k \tag{6.17}$$

The first term comes from the denominator (a recursive algorithm!) whereas the other two terms come from the numerator.

Thiran [12] and Välimäki [14] quote for the maximally flat filter for $N = 1$ the condition for the coefficient a_1:

$$a_1 = -\frac{d}{d + 2}. \tag{6.18}$$

The group delay can now be written as:

$$\tau_{\text{group}} = \Delta t + 2\frac{d\phi_D}{d\omega}. \tag{6.19}$$

The first term comes from z^{-1} whereas the second term comes from the quotient $D_N(z)/D_N(z^{-1})$.

The real and imaginary parts of the phase ϕ_D are:

$$\begin{aligned} \text{Re}\{\phi_D\} &= 1 + a_1 \cos \omega \Delta t \\ \text{Im}\{\phi_D\} &= -a_1 \sin \omega \Delta t. \end{aligned} \tag{6.20}$$

After a little tedious algebra we finally get:

$$\frac{d\phi_D}{d\omega} = -\frac{\Delta t a_1(a_1 + \cos \omega \Delta t)}{1 + a_1^2 + 2a_1 \cos \omega \Delta t}. \tag{6.21}$$

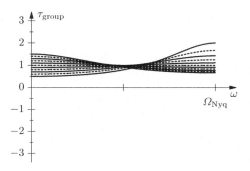

Fig. 6.4 The group delay τ_{group} in units of Δt for the $N = 1$ Thiran all-pass versus ω for delays d in the range from $-1/2$ to $+1/2$

After inserting (6.18) and using (6.19) we get:

$$\tau_{\text{group}} = \frac{\Delta t (1 + d)}{1 + d(d + 2) \sin^2 \frac{\omega \Delta t}{2}}. \tag{6.22}$$

This function is plotted in Fig. 6.4 versus ω for various values of d with $\Delta = 1$.

For $\omega \to 0$ we get $\tau_{\text{group}} = 1 + d$ (in units of Δt). No wonder, even for $d = 0$ we get 1 tag delay because we have used an *output* from the past!

Here, a word about the useful range of d is necessary. Välimäki [14] recommends $-1/2 \le d \le +1/2$ (in units of Δt). Don't worry about negative delays, we just go back to the past by half a tag and τ_{group} is always positive.

For $d = 0$ we get $\tau_{\text{group}} = 1$, for $d = 1/2$ we get $\tau_{\text{group}} = 3/2/(1 + 5/4 \sin^2(\omega \Delta t/2))$, and for $d = -1/2$ we get $\tau_{\text{group}} = 1/2(1 - 3/4 \sin^2(\omega \Delta t/2))$, all in units of Δt. For Ω_{Nyq} we get $\tau_{\text{group}} = (1+d)/(1+d(d+2))$, i.e. $\tau_{\text{group}} = 2/3$ and 2 for $d = 1/2$ and $d = -1/2$, respectively. Note the asymmetry between positive and negative delays. It also shows that negative delays are giving worse results than positive ones. In any case, there is no divergence like in the Lagrange $N = 1$ case.

We now give a few illustrations of the filter action.

6.4.1 Impulse Response

Let us see what is the response of the $N = 1$ Thiran all-pass filter to an impulse. From (6.17) we see that the output is:

$$\begin{aligned} y_0 &= a_1 \\ y_k &= \left(1 - a_1^2\right)(-a_1)^{k-1}, \quad k = 1, 2, 3, \ldots . \end{aligned} \tag{6.23}$$

This means, that—apart from the transient y_0—the output decays with increasing powers of a_1. For positive a_1 the sign alternates whereas for negative a_1 the output is always positive.

Now we show that the sum over all output data is unity, as was the input:

$$\sum_{k=0}^{\infty} y_k = a_1 + (1 - a_1^2) \sum_{k=0}^{\infty} (-a_1)^k. \tag{6.24}$$

The sum is easily evaluated with [8, No. 0.231] yielding finally:

$$\sum_{k=0}^{\infty} y_k = a_1 + \frac{1 - a_1^2}{1 + a_1} = a_1 + \frac{(1 - a_1)(1 + a_1)}{1 + a_1} = 1. \tag{6.25}$$

6.4.2 Step Response

Let us see what is the response of the $N = 1$ Thiran all-pass filter to a step, the Heaviside step function which is zero for negative times and unity at $t = 0$ and afterwards. The first example is for $d = 1/2$. Figure 6.5 illustrates the step response.

To start with, there is an undershoot of magnitude $a_1 = -1/5$, followed by rational numbers with powers of $|a_1|$ in the denominator and approaching unity monotonously.

The step response for $d = -1/2$ is a little surprising at first glance. Figure 6.6 illustrates what happens.

Fig. 6.5 The step response of the Thiran all-pass for $d = 1/2$

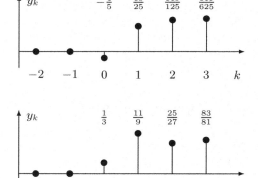

Fig. 6.6 The step response of the Thiran all-pass for $d = -1/2$

To start with, there is a positive output with $a_1 = 1/3$, followed by rational numbers with powers of a_1 in the denominator and approaching unity, but now in an oscillatory manner.

This deserves a closer look.

Apart from the transient (the first output in the present case) the following explicit formula can be derived from the difference equation (6.17).

$$y_k = 1 + (-1)^{k+1} a_1^k (1 - a_1), \qquad k = 1, 2, 3, \ldots . \tag{6.26}$$

Now it is easy to see why for negative a_1 we have a monotonous approach to unity: the prefactor of the second term is always negative because we can write:

$$(-1)^{k+1} a_1^k = (-1)^{2k+1} |a_1| \tag{6.27}$$

and the exponent is odd.

On the other hand, for positive a_1 we retain the factor $(-1)^{k+1}$ which leads to oscillations. Note that $(1 - a_1)$ is always positive in the allowed range of d.

The unit step contains all possible frequencies and the step response is what the group delay does to all those frequency components. The integral delay action is not immediately evident.

6.4.3 Ramp Response

More illustrative is the response of the Thiran all-pass with $N = 1$ to a ramp defined as:

$$f_k = \begin{cases} k & \text{for } k = 0, 1, 2, 3, \ldots \\ 0 & \text{for } k < 0 \end{cases} . \tag{6.28}$$

We should not bother that the f_k grow indefinitely. We can always stop for a given k_{max} and we are interested only in smaller k, i.e. early times such that the termination plays no role. If we normalize the f_k to the largest value, i.e. $f_{max} = 1$, we see that we actually discuss the low frequency limit.

Figures 6.7 and 6.8 show the ramp response for $d = 1/2$ and $d = -1/2$, respectively.

It is clear that in both cases the $45°$ slope is approached rapidly (monotonously for $d = 1/2$ and oscillatory for $d = -1/2$). What is very pleasing is that you can directly read off the low frequency limit of the group delay $\tau_{group} = 1 + d$, i.e. $3/2$ for $d = 1/2$ and $1/2$ for $d = -1/2$.

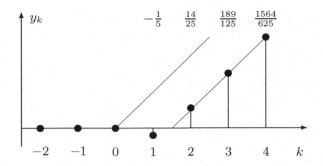

Fig. 6.7 The ramp response of the Thiran all-pass filter for $d = 1/2$; the *left line* corresponds to the ramp input whereas the *right line* indicates the input delayed by $3/2$

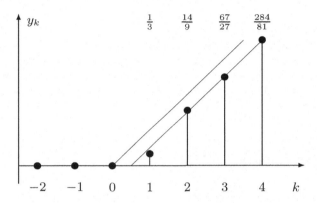

Fig. 6.8 The ramp response of the Thiran all-pass filter for $d = -1/2$; the *left line* corresponds to the ramp input whereas the *right line* indicates the input delayed by $1/2$

Again, apart from the transient, an explicit formula for y_k can be derived from the difference equation (6.17). This time, the f_k grow and the algebra is a little more tedious. Apart from the transient we arrive at the following formula:

$$y_k = ka_1 + (1 - a_1^2) \sum_{i=1}^{k-1} (-1)^{i-1} a_1^{i-1} (k - i). \tag{6.29}$$

This can be rewritten as follows:

$$y_k = ka_1 + (1 - a_1^2)(-a_1)^{k-1} \sum_{i=1}^{k-1} \frac{i}{(-a_1)^i}. \tag{6.30}$$

Amazingly enough, this sum has be evaluated in the pre-computer era yielding [8, No. 0.113]:

$$\sum_{i=1}^{k-1} \frac{i}{(-a_1)^i} = -\frac{(k-1)\left(-\frac{1}{a_1}\right)^k}{1-\left(-\frac{1}{a_1}\right)} + \left(-\frac{1}{a_1}\right)\left(\frac{1-\left(-\frac{1}{a_1}\right)^{k-1}}{\left(1-\left(-\frac{1}{a_1}\right)\right)^2}\right). \tag{6.31}$$

We now get from (6.31) after a lengthy simplification:

$$y_k = k - \left(1 - (-a_1)^k\right)\frac{1-a_1}{1+a_1}. \tag{6.32}$$

Finally, we replace a_1 by d and obtain:

$$y_k = k - (1+d)\left(1 - \left(\frac{d}{d+2}\right)^k\right). \tag{6.33}$$

For large k we get approximately:

$$y_k = k - (1+d) \tag{6.34}$$

from which we see immediately that y_k grows like k but with a delay of $1 + d$. We also see that the approach to the ramp is monotonous for positive d (always above the ramp values shifted by the delay, apart from the transient) whereas is oscillates for negative d (actually oscillating around the ramp values shifted by the delay).

It was worth while going through all the tedious algebra because we directly see the low frequency limit of $\tau_{\text{group}} = 1 + d$ in the output y_k.

Of course, in real applications, larger values for N are in use. The coefficients a_i are different, but the principle is the same. Of course, using N "past" data means that the inevitable delay will be N. If you happen to hear the same radio program with two different receivers, say on AM and FM, and one transmission sounds like "echoing" the other one, you can be sure that digital audio processing was used.

Playground

6.1 What's Your Average?
Calculate the average of the function with the Nyquist frequency and the triangular envelope as a function of the delay d of the Lagrange interpolator with $N = 1$.

6.2 Late Impulse
Calculate and plot the response of the $N = 1$ Thiran all-pass filter to an impulse for $d = 1/2$ and $d = -1/2$.

6.3 The Devil Takes the Hindmost

a. Calculate and plot the response of the $N = 1$ Thiran all-pass filter to the trailing edge of the unit pulse for $d = 1/2$.

b. Show that the sum of all output data of a unit pulse is conserved.

c. Explain why this must be the case.

6.4 Delayed Nyquist

Calculate and plot the response of the $N = 1$ Thiran all-pass for an input with $\cos \Omega_{\mathrm{Nyq}}$ starting at $t = 0$ and being 0 at earlier times for $d = 1/2$. Compare the result with the Lagrange $N = 1$ interpolator.

Chapter 7
Tomography: Backprojection of Filtered Projections

Abstract In this chapter the backprojection of filtered projections, a common algorithm in tomography, is presented. A simple example, the uniform disc, is discussed in detail in its continuous version in order to illustrate how the backprojection serves both purposes: to reconstruct the object and to "erase the shadow" where there is no object. It is shown how in a discrete version the backprojection of filtered projections is directly obtained by a convolution of the inverse Fourier transformed ramp function and the projections.

Tomography is a method to obtain 3D-images from objects without the need to cut the object into sections (like the greek word "tomos" would suggest). It would not be applicable for living objects anyway. For tomography one needs particles like X-ray photons, neutrons, high energy protons (for small objects) or even electrons (for extremely small objects) which traverse the object and which are detected behind the object. The contrast can, e.g., be given by absorption, i.e. not all particles arrive at the detector, or the energy loss of the particles. Here, we discuss only the parallel beam geometry. It could be a collimated or focussed beam (pencil geometry) which is scanned over the object or the object is moved in front of the beam, or one can use a broad beam and a position sensitive detector.

In the following we shall speak of "screen" for the detector and the result on the screen is called "projection". In order to obtain all information for the reconstruction of the 3D-image, the object has to be rotated in front of the beam, i.e. we need as many different projections under different angles as possible. In the following we shall discuss a single slice only, i.e. a 2D-object (with finite thickness, if you want). The task is to reconstruct the 2D-image from the projections. The 3D-image is then obtained by stacking the slices. Finally, slices or cross-sections in any desired direction can be computed from the full 3D-image of the object.

7.1 Projection

The projection is defined as the path integral through the object, as illustrated in Fig. 7.1. This path integral is known as the Radon transform in order to honour the person who first showed how the object can be recovered from projections. It takes a

© Springer International Publishing Switzerland 2015   173
T. Butz, *Fourier Transformation for Pedestrians*, Undergraduate Lecture
Notes in Physics, DOI 10.1007/978-3-319-16985-9_7

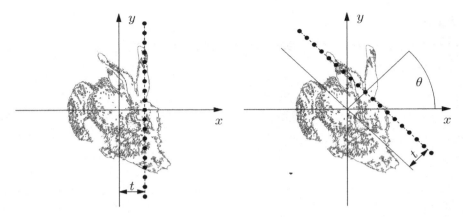

Fig. 7.1 Illustration of the path through the object for the pencil geometry. Beam parallel to y-axis (*left*); beam rotated by θ (*right*). The distance of the path from the origin is $t = x \cos \theta + y \sin \theta$

while to get used to the word "transform". We assume for simplicity that the direction of the path plays no role, i.e. it can be carried out from "front" to "back" or vice versa. Hence, we require projections for an angular range of $0 \leq \theta \leq 180°$ only. This is true only if the object does not have large density inhomogeneities. A front-only bullet-proof vest protects against a shot aiming at your breast but not when the shot hits your back!

The definition of the projection $P_\theta(t)$ of the object $\rho(x, y)$ from different angles θ is:

$$P_\theta(t) = \int\limits_{-\infty}^{+\infty} \rho(x, y)\, \delta(x \cos \theta + y \sin \theta - t)\, \mathrm{d}x\mathrm{d}y \qquad (7.1)$$

Here, the δ-"function" is unity if and only if its argument is 0. In other words, the δ-"function" ensures that only points (x, y) contribute to the integral which are on a line with a distance t to the origin. We could rotate the sample by $-\theta$ in Fig. 7.1(*right*) in order to leave the beam vertical. Fortunately, we are free to rotate the sample clockwise or counter-clockwise, so we do not have to worry about the sign of θ. You may look up [8, No. 3.02.2] and see that the sign of θ does not matter in the integration over θ.

Stacking all $P_\theta(t)$ versus θ we get what is called a "sinogram" because a single mass point rotating in front of the screen would give a sinusoidal projection, eventually with a phase. Since the object could be considered to consist of mass points (maybe better voxels) the sinogram is a superposition of the sinusoidal projections of all mass points. Thus the sinogram contains all necessary information for the image reconstruction.

7.2 Backprojection of Filtered Projections

A standard method for the image reconstruction in tomography is the so-called "Backprojection of filtered projections". Needless to say that we require a Fourier transformation to illustrate the method. Here, we use a spatial coordinate x and an "angular wave number" k instead of time t and angular frequency ω as we did thus far. A simple periodic density, e.g., will be written like

$$\rho(x) = \cos kx. \tag{7.2}$$

It is immediately clear that a simple direct backprojection of the projection would not do the job. We would get density in the space between the object and the screen which would be wrong. If there is no density between the object and the screen, the projection would not depend on the distance between object and screen at all for parallel beams. Hence, what matters is the change of the projection with respect to the beam direction, i.e. its spatial derivative. This is also what a rigorous treatment of the problem would tell us. In order to avoid the awesome view of a screen penetrating the object it is helpful to consider a single mass point within the object. The derivative of the projection would be 0 in front of the object and would suddenly switch to high as the voxel is touched. We could do so for all voxels in the slice and—due to linearity—finally superimpose all results. What we get are streaks within the object space proportional to the path integral trough the object and aligned with the path. The superposition of all streaky patterns for all angles θ finally leads to the desired 2D-density distribution.

Now, rather than to do the inverse Radon transform, the Fourier transformation comes into play:

$$S_\theta(k) = \int\limits_{-\infty}^{+\infty} P_\theta(t) \mathrm{e}^{-ikt} \mathrm{d}t. \tag{7.3}$$

The Fourier transformation is carried out in 1D only because of the Fourier slice theorem, sometimes also called the projection-slice theorem or central slice theorem. It states that the 1D-Fourier transform of a 1D-projection of a 2D-object is the same as extracting a central slice—better a path—through the origin along $k_y = 0$ of the 2D-Fourier transform of the object. You are encouraged to proof it in the exercises.

It is annoying that the word "slice" in the Fourier slice theorem has a different meaning than the slice we were speaking about when stacking slices of 2D-images in order to get a 3D-reconstruction.

Now we remember that the Fourier transform of the derivative of a function is simply the Fourier transform of the function multiplied by $i\omega$ (see (2.4)), or ik in the present case. Hence the Fourier transform of the derivative of the projection is:

$$P_\theta(t) = ik S_\theta(k). \tag{7.4}$$

Since the direction of the path in the path integral (7.1) plays no role, we can use $|ik| = |k|$ as the multiplier. This multiplier is usually called the "ramp filter". It is a linear high-pass. For very long wavelength it practically suppresses $S_\theta(k)$ whereas for high frequencies it enhances $S_\theta(k)$.

Next, we carry out the inverse Fourier transform:

$$Q_\theta(t) = \frac{1}{(2\pi)^2} \int\limits_{-\infty}^{+\infty} S_\theta(k)|k|e^{+ikt}dk. \tag{7.5}$$

I am sure you have noticed the prefactor $\frac{1}{(2\pi)^2}$. Where does it come from? Let's argue with the consideration of units. For the sake of simplicity we have used no units for $\rho(x)$. Hence, the projection $P_\theta(t)$ has the unit "length" and $S_\theta(k)$ the unit "area". The inverse Fourier transform is carried out by the integral of (7.5) and the angular integral (7.6). Hence, the integral in (7.5) accounts for two dimensions and thus we require $\frac{1}{(2\pi)^2}$ as prefactor. At this point I am tempted to regret not to have used 2π times the wave number instead of the angular wave number (like $2\pi\nu$ instead of ω) because there would be no prefactor at all, not in one nor in two dimensions. However, we would have missed the insight into the secrets of what mathematicians call "cylinder coordinates".

Finally, we have to integrate over all angles θ:

$$\rho(x) = \int\limits_{0}^{\pi} Q_\theta(x\cos\theta + y\sin\theta)\,d\theta. \tag{7.6}$$

Here, we have used $t = x\cos\theta + y\sin\theta$. The last two steps represent the "backprojection".

This procedure looks very complicated, and in reality, nobody would follow this recipe, as we shall see at the end of this chapter. However, it is instructive to see, how "density" is backprojected where we want it and how it is "erased" where we don't.

Let us do a "simple" example:

Example 7.1 ("Spherical disk" of uniform density)

$$\rho(r) = \begin{cases} 1 \text{ for } r \leq R \\ 0 \text{ else} \end{cases}. \tag{7.7}$$

Due to symmetry it suffices to consider $\rho(x)$ only. In this case all $P_\theta(t)$ are identical and the integration over θ should be trivial. This is true for $t \leq R$, but not so for the "outside", i.e. t larger than R, as we shall see!

The projection reads:

$$P_\theta(t) = \begin{cases} 2R\sqrt{1 - \left(\frac{t}{R}\right)^2} & \text{inside} \\ 0 & \text{outside} \end{cases}. \qquad (7.8)$$

The Fourier transform of it reads:

$$S_\theta(k) = 2 \int_0^R 2R\sqrt{1 - \left(\frac{t}{R}\right)^2} \cos kt \, dt = 2\pi R^2 \frac{J_1(kR)}{kR}. \qquad (7.9)$$

Here, we have used [8, No. 3.752.2]. $J_1(kR)$ is a Bessel function, no surprise for cylindrical symmetry. Don't worry about the denominator. The $J_1(kR)$ goes to 0 with the same power of k thus the expression remains finite for $k = 0$, see Fig. 7.2.

The factor of 2 in front of the integral comes from the fact that we integrate over positive t only. This is sufficient for an even integrand. Similarly, we do not have to bother about the imaginary part of the Fourier transform, it simply vanishes.

We also see that $S_\theta(k)$ is an even function because the Bessel function $J_1(kR)$ is odd as is the denominator.

Fig. 7.2 $J_1(kR)/kR$ versus k for $R = 1$

Then we calculate with [15, 541 3b]:

$$
Q_\theta(t) = \begin{cases} \dfrac{1}{\pi} & \text{for } t < R \\[3mm] -\dfrac{1}{\pi}\left(\dfrac{1}{\sqrt{1-\left(\frac{R}{t}\right)^2}} - 1 \right) & \text{for } t \geq R \end{cases} . \tag{7.10}
$$

We note that for t "inside R" $Q_\theta(t)$ does not depend on t. Integrating $Q_\theta(t)$ from 0 to π just gives $\rho(x) = 1$ or $\rho(r) = 1$, as it should be. Thus far for the "painting" part of the backprojection.

Surprisingly, at first glance, $Q_\theta(t)$ for t "outside R" is negative and even diverges as t approaches R. No wonder, this is our "rubber" which erases everything outside. How this is done requires the integration over $Q_\theta(t)$ of an arc from $\theta = 0$ to $\arccos(R/x)$ with the "outside" integrand and from $\theta = \arccos(R/x)$ to $\pi/2$ with the "inside" integrand, as illustrated in Fig. 7.3.

$$
\rho(x) = 2 \left(\int_0^{\arccos(R/x)} -\frac{1}{\pi}\left(\frac{1}{\sqrt{1-\left(\frac{R}{t}\right)^2}} - 1 \right) d\theta + \int_{\arccos(R/x)}^{\pi/2} \frac{1}{\pi} \, d\theta \right)
$$

$$
= 2 \left(\frac{1}{\pi} \int_0^{\pi/2} d\theta - \frac{1}{\pi} \int_0^{\arccos(R/x)} \frac{1}{\sqrt{1-\left(\frac{R}{x\cos^2\theta}\right)^2}} \, d\theta \right) \tag{7.11}
$$

The factor 2 in front comes from the fact that it suffices to integrate up to $\pi/2$ only due to symmetry. In the last step we have used $t = x\cos\theta$ because it suffices to calculate ρ along the x-axis only. We can later replace x by r, again by symmetry.

The first integral is trivial and yields 1, including the factor of 2 in front. The second integral is a little tricky. We substitute $t = \tan\theta$, $d\theta = \left(1/\left(1+t^2\right)\right) dt$, and $1/\cos^2\theta = 1+t^2$, as suggested in [16, p. 117].

After some rearrangements we end up with the following expression for the second integral including the prefactor 2:

$$
-\frac{2}{\pi} \arctan\left(\frac{t}{\sqrt{\left(\frac{x}{R}\right)^2 - 1 - t^2}} \right) \Bigg|_0^{\sqrt{\left(\frac{x}{R}\right)^2 - 1}} . \tag{7.12}
$$

Here we used [16, 20, 236]. Inserting the upper boundary we see that the denominator goes to 0. Hence, we finally obtain -1 because $\arctan\infty = \pi/2$ and the lower boundary does not contribute anything. Hence, we finally get $\rho(x) = 0$ outside, as required. Due to rotational symmetry we can also write $\rho(r) = 0$ outside.

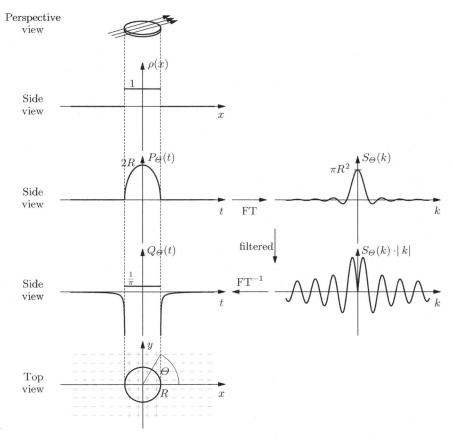

Fig. 7.3 Illustration of the "recipe" from *top* to *bottom*: projection of homogeneous disc $P_\theta(t)$; its Fourier transform $S_\theta(k)$; $Q_\theta(t)$, the inverse Fourier transform of $S_\theta(k)$ multiplied by the ramp $|k|$; integration of $Q_\theta(t)$ over θ for t "outside R"

We see that the expression for $Q_\theta(t)$ with t "outside R" integrated over the corresponding arc exactly "erases" what the expression for $Q_\theta(t)$ with t "inside R" with its corresponding complementary arc has "painted".

Since the length of the arc outside tends to 0 the closer we come with t to R, the "rubber" must be infinitely efficient. Now we understand the singularity in (7.10).

All the above individual steps are illustrated in Fig. 7.3.

At the end we show, as promised, how the backprojection of filtered projections is carried out in reality. The whole effort in Fourier transforming the projection, weighting, and backtransforming is completely unnecessary using the convolution theorem: the integrand of the backtransformation is the product of the Fourier-transformed projection $S_\theta(k)$ and the "ramp" $|k|$. This is identical to the convolution of the projection itself (forth and back transformation compensate each other) with the inverse Fourier transform of the "ramp".

Now you might wonder why we did not do it like this right away. The answer is: we cannot! The inverse Fourier transform of $|k|$ does not exist, the integral diverges. Don't worry! In the discrete world we have a finite length Δx between neighbouring data points of the projection and, hence, a k_{Nyq}. This corresponds to pixels alternating between 1 and 0. Faster variations of the projection data simply do not exist. Hence, the inverse Fourier transform of the finite ramp exists, of course.

We have already calculated something very similar to the inverse Fourier transform of the double-sided ramp, i.e. the Fourier transform of the triangular function. The positive sign in the exponent does not matter since we are dealing with an even function. What should not be forgotten is that there is no prefactor $\frac{1}{2M}$ for the inverse Fourier transform. I leave it to the exercise to calculate it.

Thus a discrete convolution is not very time consuming. What remains is the summation over the discrete angles θ.

In practice, of course, more sophisticated equipment like fan geometry beams or even 3D-detectors in positron annihilation tomography (PET) are used as well as other "filters". Often more sophisticated reconstruction algorithms like iterative algorithms are required which could cope with complications like lateral straggling of the beam, absorption of soft X-rays or a limited range of projection angles. We pedestrians are happy with the basics.

Playground

7.1 Go on the Ramp, not on the Rampage!
Calculate the inverse discrete Fourier transform of the discrete double-sided ramp.

Hint: You may either use the first shifting rule or remember the tricky Carl Friedrich Gauß.

7.2 Slice It!
Proof the Fourier slice theorem.

Hint: Let the projection be onto the x-axis. There is no loss of generality because we are free to choose the coordinate system.

7.3 Reconstruct It!
Suppose, we have the following object with two projections (smallest, non-trivial symmetric image):

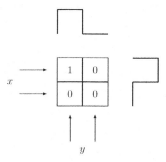

If it helps, consider a cube of uniform density and its shadow (=projection) when illuminated with a light-beam from the x- and y-direction. 1 = there is a cube, 0 = there is no cube (but here we have a 2D-problem).

Use a ramp filter, defined as $\{G_0 = 0, G_1 = 1\}$ and periodic continuation in order to convolve the projection with the Fourier-transformed ramp-filter and project the filtered data back. Discuss all possible different images.

Hint: Perform convolution along the x- and y-direction consecutively.

Appendix: Solutions

Playground of Chap. 1

1.1 Very Speedy

$$\omega = 2\pi\nu \quad \text{with } \nu = 100 \times 10^6 \, \text{s}^{-1}$$
$$= 628.3 \, \text{Mrad/s}$$
$$T = \frac{1}{\nu} = 10 \, \text{ns} \, ; \, s = cT = 3 \times 10^8 \, \text{m/s} \times 10^{-8} \, \text{s} = 3 \, \text{m}.$$

Easy to remember: 1 ns corresponds to 30 cm, the length of a ruler.

1.2 Totally Odd

It is mixed since neither $f(t) = f(-t)$ nor $f(-t) = -f(t)$ is true. The graphical solution is shown in Fig. A.1. The Fourier series for both $f_{\text{even}}(t)$ and $f_{\text{odd}}(t)$ are infinite series; the series for the even part decreases with $1/k^2$ (kink) whereas that for the odd part decreases with $1/k$ (discontinuity).

1.3 Absolutely True

This is an even function! It could have been written as $f(t) = |\sin \pi t|$ in $-\infty \leq t \leq +\infty$ as well. It is most convenient to integrate from 0 to 1, i.e. a full period of unit length.

$$C_k = \int_0^1 \sin \pi t \cos 2\pi k t \, dt$$

$$= \int_0^1 \frac{1}{2} [\sin(\pi - 2\pi k)t + \sin(\pi + 2\pi k)t] \, dt$$

© Springer International Publishing Switzerland 2015
T. Butz, *Fourier Transformation for Pedestrians*, Undergraduate Lecture
Notes in Physics, DOI 10.1007/978-3-319-16985-9

$$= \frac{1}{2}\left\{(-1)\frac{\cos(\pi - 2\pi k)t}{\pi - 2\pi k}\bigg|_0^1 + (-1)\frac{\cos(\pi + 2\pi k)t}{\pi + 2\pi k}\bigg|_0^1\right\}$$

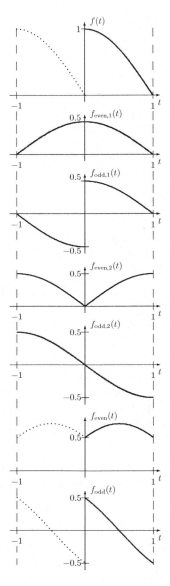

Fig. A.1 $f(x) = \cos(\pi t/2)$ for $0 \le t \le 1$, periodic continuation in the interval $-1 \le t \le 0$ is *dotted*; the following two graphs add up correctly for the interval $0 \le t \le 1$ but give 0 for the interval $-1 \le t \le 0$; the next two graphs add up correctly for the interval $-1 \le t \le 0$ and leave the interval $0 \le t \le 1$ unchanged; the bottom two graphs show $f_{\text{even}}(t) = f_{\text{even},1}(t) + f_{\text{even},2}(t)$ and $f_{\text{odd}}(t) = f_{\text{odd},1}(t) + f_{\text{odd},2}(t)$ (*from top to bottom*)

$$= \frac{1}{2} \left\{ \frac{(-1)\cos \pi (1-2k)}{\pi - 2\pi k} + \frac{1}{\pi - 2\pi k} + \frac{(-1)\cos \pi (1+2k)}{\pi + 2\pi k} + \frac{1}{\pi + 2\pi k} \right\}$$

$$= \frac{1}{2} \left\{ (-1) \left[\frac{\overset{=(-1)}{\cos \pi} \overset{=1}{\cos 2\pi k} + \overset{=0}{\sin \pi} \sin 2\pi k}{\pi - 2\pi k} \right] \right.$$

$$\left. + (-1) \left[\frac{\overset{=(-1)}{\cos \pi} \overset{=1}{\cos 2\pi k} - \overset{=0}{\sin \pi} \sin 2\pi k}{\pi + 2\pi k} \right] + \frac{2\pi}{\pi^2 - 4\pi^2 k^2} \right\}$$

$$= \frac{1}{2} \left\{ \frac{1}{\pi - 2\pi k} + \frac{1}{\pi + 2\pi k} + \frac{2\pi}{\pi^2 - 4\pi^2 k^2} \right\}$$

$$= \frac{2}{\pi - 4\pi k^2} \quad = \frac{2}{\pi (1 - 4k^2)}$$

$$f(t) = \overset{k=0}{\frac{2}{\pi}} - \overset{k=\pm 1}{\frac{4}{3\pi} \cos 2\pi t} - \overset{k=\pm 2}{\frac{4}{15\pi} \cos 4\pi t} - \overset{k=\pm 3}{\frac{4}{35\pi} \cos 6\pi t} - \cdots .$$

1.4 Rather Complex

The function $f(t) = 2 \sin(3\pi t/2) \cos(\pi t/2)$ for $0 \le t \le 1$ can be rewritten using a trigonometric identity as $f(t) = \sin \pi t + \sin 2\pi t$. We have just calculated the first part and the linearity theorem tells us that we only have to calculate C_k for the second part and then add both coefficients. The second part is an odd function! We actually do not have to calculate C_k because the second part is our basis function for $k = 1$. Hence,

$$C_k = \begin{cases} i/2 & \text{for } k = +1 \\ -i/2 & \text{for } k = -1 \\ 0 & \text{else} \end{cases} .$$

Together:

$$C_k = \frac{2}{\pi (1 - 4k^2)} + \frac{i}{2} \delta_{k,1} - \frac{i}{2} \delta_{k,-1} .$$

The graphical solution is displayed in Fig. A.2.

1.5 Shiftily

With the First Shifting Rule we get:

$$C_k^{\text{new}} = e^{+i2\pi k \frac{1}{2}} C_k^{\text{old}}$$
$$= e^{+i\pi k} C_k^{\text{old}} = (-1)^k C_k^{\text{old}} .$$

Shifted first part:
even terms remain unchanged, odd terms get a minus sign. We would have to calculate:

$$C_k = \int\limits_{-1/2}^{1/2} \cos \pi t \cos 2\pi k t \; dt.$$

Shifted second part:
imaginary parts for $k = \pm 1$ now get a minus sign because the amplitude is negative.

Figure A.3 illustrates both shifted parts. Note the kink at the center of the interval which results from the fact that the slopes of the unshifted function at the interval boundaries are different (see Fig. A.2).

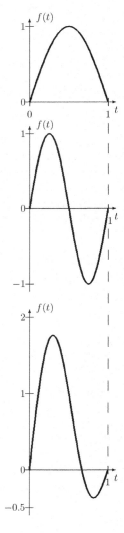

Fig. A.2 $\sin \pi t$ (*top*); $\sin 2\pi t$ (*middle*); sum of both (*bottom*)

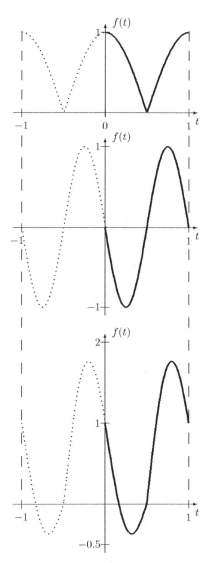

Fig. A.3 Shifted first part, shifted second part, sum of both (*from top to bottom*)

1.6 Cubed

The function is even, the C_k are real. With the trigonometric identity $\cos^3 2\pi t = (1/4)(3\cos 2\pi t + \cos 6\pi t)$ we get:

$$
\begin{array}{ccc}
C_0 = 0 & & A_0 = 0 \\
C_1 = C_{-1} = 3/8 & \text{or} & A_1 = 3/4. \\
C_3 = C_{-3} = 1/8 & & A_3 = 1/4
\end{array}
$$

Check using the Second Shifting Rule: $\cos^3 2\pi t = \cos 2\pi t \cos^2 2\pi t$. From (1.5) we get $\cos^2 2\pi t = 1/2 + (1/2)\cos 4\pi t$, i.e. $C_0^{\text{old}} = 1/2$, $C_2^{\text{old}} = C_{-2}^{\text{old}} = 1/4$.

From (1.36) with $T = 1$ and $a = 1$ we get for the real part (the B_k are 0):

$$C_0 = A_0; \quad C_k = A_{k/2}; \quad C_{-k} = A_{k/2},$$

$$C_0^{\text{old}} = 1/2 \quad \text{and} \quad C_2^{\text{old}} = C_{-2}^{\text{old}} = 1/4$$

with $C_k^{\text{new}} = C_{k-1}^{\text{old}}$:

$$
\begin{aligned}
C_0^{\text{new}} &= C_{-1}^{\text{old}} = 0 \\
C_1^{\text{new}} &= C_0^{\text{old}} = 1/2 \qquad && C_{-1}^{\text{new}} = C_{-2}^{\text{old}} = 1/4 \\
C_2^{\text{new}} &= C_1^{\text{old}} = 0 \qquad && C_{-2}^{\text{new}} = C_{-3}^{\text{old}} = 0 \\
C_3^{\text{new}} &= C_2^{\text{old}} = 1/4 \qquad && C_{-3}^{\text{new}} = C_{-4}^{\text{old}} = 0.
\end{aligned}
$$

Note, that for the shifted C_k we do no longer have $C_k = C_{-k}$! Let us construct the A_k^{new} first:

$$A_k^{\text{new}} = C_k^{\text{new}} + C_{-k}^{\text{new}}$$

$A_0^{\text{new}} = 0$; $A_1^{\text{new}} = 3/4$; $A_2^{\text{new}} = 0$; $A_3^{\text{new}} = 1/4$. In fact, we want to have $C_k = C_{-k}$, so we better define $C_0^{\text{new}} = A_0^{\text{new}}$ and $C_k^{\text{new}} = C_{-k}^{\text{new}} = A_k^{\text{new}}/2$.

Figure A.4 shows the decomposition of the function $f(t) = \cos^3 2\pi t$ using a trigonometric identity.

The Fourier coefficients C_k of $\cos^2 2\pi t$ before and after shifting using the Second Shifting Rule as well as the Fourier coefficients A_k for $\cos^2 2\pi t$ and $\cos^3 2\pi t$ are displayed in Fig. A.5.

1.7 Tackling Infinity

Let $T = 1$ and set $B_k = 0$. Then we have from (1.50):

$$\int_0^1 f(t)^2 \mathrm{d}t = A_0^2 + \frac{1}{2}\sum_{k=1}^{\infty} A_k^2.$$

We want to have $A_k^2 \propto 1/k^4$ or $A_k \propto \pm 1/k^2$. Hence, we need a kink in our function, like in the "triangular function". However, we do not want the restriction to odd k. Let's try a parabola. $f(t) = t(1 - t)$ for $0 \le t \le 1$.

Fig. A.4 The function $f(t) = \cos^3 2\pi t$ can be decomposed into $f(t) = (3\cos 2\pi t + \cos 6\pi t)/4$ using a rigonometric identity

For $k \neq 0$ we get:

$$C_k = \int\limits_0^1 t(1-t)\cos 2\pi kt\, \mathrm{d}t$$

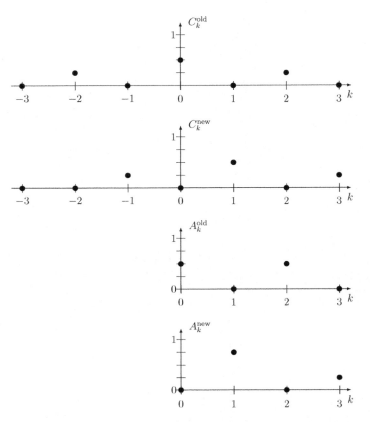

Fig. A.5 Fourier coefficients C_k for $f(t) = \cos^2 2\pi t = 1/2 + (1/2)\cos 4\pi t$ and after shifting using the Second Shifting Rule (*top two*). Fourier coefficients A_k for $f(t) = \cos^2 2\pi t$ and $f(t) = \cos^3 2\pi t$ (*bottom two*)

$$
= \int_0^1 t\cos 2\pi kt \, \mathrm{d}t - \int_0^1 t^2 \cos 2\pi kt \, \mathrm{d}t
$$

$$
= \frac{\cos 2\pi kt}{(2\pi k)^2}\bigg|_0^1 + \frac{t\sin 2\pi kt}{2\pi k}\bigg|_0^1
$$

$$
- \left(\frac{2t}{(2\pi k)^2}\cos 2\pi kt + \left(\frac{t^2}{2\pi k} - \frac{2}{(2\pi k)^3} \right)\sin 2\pi kt \right)\bigg|_0^1
$$

$$
= -\left(\frac{2}{(2\pi k)^2} \times 1 + \left(\frac{1}{2\pi k} - \frac{2}{(2\pi k)^3} \right) \times 0 - \left(0 - \frac{2}{(2\pi k)^3} \right) \times 0 \right)
$$

$$
= -\frac{1}{2\pi^2 k^2}.
$$

For $k = 0$ we get:

$$C_0 = \int_0^1 t(1-t)\mathrm{d}t = \int_0^1 t\mathrm{d}t - \int_0^1 t^2\mathrm{d}t$$

$$= \frac{t^2}{2}\Big|_0^1 - \frac{t^3}{3}\Big|_0^1 = \frac{1}{2} - \frac{1}{3}$$

$$= \frac{1}{6}.$$

From the left hand side of (1.50) we get:

$$\int_0^1 t^2(1-t)^2\mathrm{d}t = \int_0^1 (t^2 - 2t^3 + t^4)\mathrm{d}t$$

$$= \frac{t^3}{3} - 2\frac{t^4}{4} + \frac{t^5}{5}\Big|_0^1 = \frac{1}{3} - \frac{1}{2} + \frac{1}{5}$$

$$= \frac{10 - 15 + 6}{30}$$

$$= \frac{1}{30}.$$

Hence, with $A_0 = C_0$ and $A_k = C_k + C_{-k} = 2C_k$ we get:

$$\frac{1}{30} = \frac{1}{36} + \frac{1}{2}\sum_{k=1}^{\infty}\left(\frac{1}{\pi^2 k^2}\right)^2 = \frac{1}{36} + \frac{1}{2\pi^4}\sum_{k=1}^{\infty}\frac{1}{k^4}$$

$$\text{or } \left(\frac{1}{30} - \frac{1}{36}\right)2\pi^4 = \sum_{k=1}^{\infty}\frac{1}{k^4} = \frac{36 - 30}{1080}2\pi^4$$

$$= \frac{6\pi^4}{540} = \frac{\pi^4}{90}.$$

1.8 Smoothly

From (1.63) we know that a discontinuity in the function leads to a $\left(\frac{1}{k}\right)$-dependence; a discontinuity in the first derivative leads to a $\left(\frac{1}{k^2}\right)$-dependence etc.

Here, we have:

$$
\begin{aligned}
f &= 1 - 8t^2 + 16t^4 & \text{is continuous at the boundaries} \\
f' &= -16t + 64t^3 = -16t(1 - 4t^2) & \text{is continuous at the boundaries} \\
f'' &= -16 + 192t^2 & \text{is still continuous at the boundaries} \\
f''' &= 384t & \text{is \textit{not} continuous at the boundaries} \\
f'''\left(-\tfrac{1}{2}\right) &= -192 \quad f'''\left(+\tfrac{1}{2}\right) = +192.
\end{aligned}
$$

Hence, we should have a $\left(\frac{1}{k^4}\right)$-dependence.

Check by direct calculation. For $k \neq 0$ we get:

$$C_k = \int\limits_{-1/2}^{+1/2} (1 - 8t^2 + 16t^4) \cos 2\pi kt \, dt$$

$$= 2 \int\limits_{0}^{1/2} (\cos 2\pi kt - 8t^2 \cos 2\pi kt + 16t^4 \cos 2\pi kt) dt \quad \text{with } a = 2\pi k$$

$$= 2 \left[\frac{\sin at}{a} - 8 \left[\frac{2t}{a^2} \cos at + \left(\frac{t^2}{a} - \frac{2}{a^3} \right) \sin at \right] \right.$$

$$\left. + t^4 \frac{\sin at}{a} - \frac{4}{a} \left[\left(\frac{3t^2}{a^2} - \frac{6}{a^4} \right) \sin at - \left(\frac{t^3}{a} - \frac{6t}{a^3} \right) \cos at \right] \right]\Bigg|_0^{1/2}$$

$$= 2 \left[-8 \left(\frac{1}{a^2} (-1)^k \right) + 16 \frac{1}{2a^4} (-1)^k (a^2 - 24) \right]$$

$$= 2(-1)^k \left(\frac{8}{a^2} + \frac{8}{a^4} (a^2 - 24) \right)$$

$$= 16(-1)^k \left(-\frac{1}{a^2} + \frac{1}{a^2} - \frac{24}{a^4} \right)$$

$$= -16 \times 24 \frac{(-1)^k}{a^4}$$

$$= -384 \frac{(-1)^k}{a^4}$$

$$= -\frac{24(-1)^k}{\pi^4 k^4}.$$

For $k = 0$ we get:

$$C_0 = 2 \int\limits_{0}^{1/2} (1 - 8t^2 + 16t^4) dt$$

$$= 2 \left(t - \frac{8}{3} t^3 + \frac{16}{5} t^5 \right)\Bigg|_0^{1/2}$$

$$= 2 \left(\frac{1}{2} - \frac{8}{3} \frac{1}{8} + \frac{16}{5} \frac{1}{32} \right)$$

$$= 2 \left(\frac{1}{2} - \frac{1}{3} + \frac{1}{10} \right) = 2 \frac{15 - 10 + 3}{30}$$

$$= \frac{8}{15}.$$

Playground of Chap. 2

2.1 Black Magic

Figure A.6 illustrates the construction:

i. The inclined straight line is $y = x \tan \theta$, the straight line parallel to the x-axis is $y = a$. Their intersection yields $x \tan \theta = a$ or $x = a \cot \theta$.
   The circle is written as $x^2 + (y - a/2)^2 = (a/2)^2$ or $x^2 + y^2 - ay = 0$. Inserting $x = y \cot \theta$ for the inclined straight line yields $y^2 \cot^2 \theta + y^2 = ay$ or $-$ dividing by $y \neq 0 - y = a/(1 + \cot^2 \theta) = a \sin^2 \theta$ (the trivial solution $y = 0$ corresponds to the intersection at the origin and $\pm\infty$).

ii. Eliminating θ we get $y = a/(1 + (x/a)^2) = a^3/(a^2 + x^2)$.

iii. Calculating the Fourier transform is the reverse problem of (2.17):

$$F(\omega) = 2 \int_0^\infty \frac{a^3}{a^2 + x^2} \cos \omega x \, \mathrm{d}x$$

$$= 2a^3 \int_0^\infty \frac{\cos \omega a x'}{a^2 + a^2 x'^2} a \, \mathrm{d}x' \quad \text{with } x = ax'$$

$$= 2a^2 \int_0^\infty \frac{\cos \omega a x'}{1 + x'^2} \mathrm{d}x'$$

$$= a^2 \pi e^{-a|\omega|}$$

the double-sided exponential. In fact, what mathematicians call the "versiera" of Agnesi is—apart from constants—identical to what physicists call a Lorentzian.

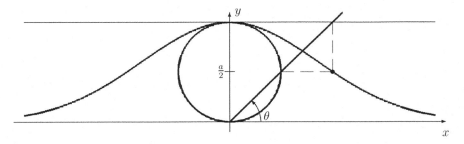

Fig. A.6 The "versiera" of Agnesi: a construction recipe for a Lorentzian with ruler and circle

What about "Black magic"? A rational function, the geometric locus of a simple problem involving straight lines and a circle, has a transcendental Fourier transform and vice versa! No surprise, the trigonometric functions used in the Fourier transformation are transcendental themselves!

2.2 The Phase Shift Knob

We write $f(t) \leftrightarrow \text{Re}\{F(\omega)\} + i\,\text{Im}\{F(\omega)\}$ before shifting. With the First Shifting Rule we get:

$$
\begin{aligned}
f(t - a) &\leftrightarrow (\text{Re}\{F(\omega)\} + i\,\text{Im}\{F(\omega)\})\,(\cos \omega a - i \sin \omega a) \\
&= \text{Re}\{F(\omega)\} \cos \omega a + \text{Im}\{F(\omega)\} \sin \omega a \\
&\quad + i\,(\text{Im}\{F(\omega)\} \cos \omega a - \text{Re}\{F(\omega)\} \sin \omega a).
\end{aligned}
$$

The imaginary part vanishes for $\tan \omega a = \text{Im}\{F(\omega)\}/\text{Re}\{F(\omega)\}$ or $a = (1/\omega) \times \arctan(\text{Im}\{F(\omega)\}/\text{Re}\{F(\omega)\})$. For a sinusoidal input with phase shift, i.e. $f(t) = \sin(\omega t - \varphi)$, we identify a with φ/ω, hence $\varphi = a \arctan(\text{Im}\{F(\omega)\}/\,\text{Re}\{F(\omega)\})$. This is our "phase shift knob". If, e.g., $\text{Re}\{F(\omega)\}$ were 0 before shifting, we would have to turn the "phase shift knob" by $\omega a = \pi/2$ or—with $\omega = 2\pi/T$—by $a = T/4$ (or $90°$, i.e. the phase shift between sine and cosine). Since $\text{Re}\{F(\omega)\}$ was non-zero before shifting, less than $90°$ is sufficient to make the imaginary part vanish. The real part which builds up upon shifting must be $\text{Re}\{F_{\text{shifted}}\} = \sqrt{\text{Re}\{F(\omega)\}^2 + \text{Im}\{F(\omega)\}^2}$ because $|F(\omega)|$ is unaffected by shifting and $\text{Im}\{F_{\text{shifted}}\} = 0$. If you are skeptic insert $\tan \omega a = \text{Im}\{F(\omega)\}/\text{Re}\{F(\omega)\}$ into the expression for $\text{Re}\{F_{\text{shifted}}\}$:

$$
\begin{aligned}
\text{Re}\{F_{\text{shifted}}\} &= \text{Re}\{F(\omega)\} \cos \omega a + \text{Im}\{F(\omega)\} \sin \omega a \\
&= \text{Re}\{F(\omega)\} \frac{1}{\sqrt{1 + \tan^2 \omega a}} + \text{Im}\{F(\omega)\} \frac{\tan \omega a}{\sqrt{1 + \tan^2 \omega a}} \\
&= \frac{\text{Re}\{F(\omega)\} + \text{Im}\{F(\omega)\}\frac{\text{Im}\{F(\omega)\}}{\text{Re}\{F(\omega)\}}}{\sqrt{1 + \frac{\text{Im}\{F(\omega)\}^2}{\text{Re}\{F(\omega)\}^2}}} \\
&= \sqrt{\text{Re}\{F(\omega)\}^2 + \text{Im}\{F(\omega)\}^2}.
\end{aligned}
$$

Of course, the "phase shift knob" does the job only for a given frequency ω.

2.3 Pulses

$f(t)$ is odd; $\omega_0 = n\frac{2\pi}{T/2}$ or $\frac{T}{2}\omega_0 = n2\pi$.

$$
F(\omega) = (-i) \int\limits_{-T/2}^{T/2} \sin(\omega_0 t) \sin \omega t \, dt
$$

$$= (-i)\frac{1}{2}\int\limits_{-T/2}^{T/2} (\cos(\omega_0 - \omega)t - \cos(\omega_0 + \omega)t)\,dt$$

$$= (-i)\int\limits_{0}^{T/2} (\cos(\omega_0 - \omega)t - \cos(\omega_0 + \omega)t)\,dt$$

$$= (-i)\left(\frac{\sin(\omega_0 - \omega)\frac{T}{2}}{\omega_0 - \omega} - \frac{\sin(\omega_0 + \omega)\frac{T}{2}}{\omega_0 + \omega}\right)$$

$$= (-i)\left(\frac{\overset{=0}{\sin\omega_0\frac{T}{2}}\,\overset{=1}{\cos\omega\frac{T}{2}} - \cos\omega_0\frac{T}{2}\,\sin\omega\frac{T}{2}}{\omega_0 - \omega}\right.$$

$$\left. - \frac{\overset{=0}{\sin\omega_0\frac{T}{2}}\,\overset{=1}{\cos\omega\frac{T}{2}} + \cos\omega_0\frac{T}{2}\,\sin\omega\frac{T}{2}}{\omega_0 + \omega}\right)$$

$$= i\sin\omega\frac{T}{2}\left(\frac{1}{\omega_0 - \omega} + \frac{1}{\omega_0 + \omega}\right) = 2i\sin\frac{\omega T}{2} \times \frac{\omega_0}{\omega_0^2 - \omega^2}.$$

At resonance: $F(\omega_0) = -iT/2$; $F(-\omega_0) = +iT/2$; $|F(\pm\omega_0)| = T/2$. This is easily seen by going back to the expressions of the type $\frac{\sin x}{x}$.

For two such pulses centered around $\pm\Delta$ we get:

$$F_{\text{shifted}}(\omega) = 2i\sin\frac{\omega T}{2} \times \frac{\omega_0}{\omega_0^2 - \omega^2}\left(e^{i\omega\Delta} + e^{-i\omega\Delta}\right)$$

$$= 4i\sin\frac{\omega T}{2} \times \frac{\omega_0}{\omega_0^2 - \omega^2}\cos\omega\Delta \quad\longleftarrow \text{"modulation"}.$$

$|F(\omega_0)| = T$ if at resonance: $\omega_0\Delta = l\pi$. In order to maximise $|F(\omega)|$ we require $\omega\Delta = l\pi$; $l = 1, 2, 3, \ldots$; Δ depends on ω!

2.4 Phase-Locked Pulses

This is a textbook case for the Second Shifting Rule! Hence, we start with DC-pulses. This function is even!

$$F_{\text{DC}}(\omega) = \int\limits_{-\Delta-\frac{T}{2}}^{-\Delta+\frac{T}{2}} \cos\omega t\,dt + \int\limits_{+\Delta-\frac{T}{2}}^{+\Delta+\frac{T}{2}} \cos\omega t\,dt = 2\int\limits_{\Delta-\frac{T}{2}}^{\Delta+\frac{T}{2}} \cos\omega t\,dt$$

with $t' = -t$ we get a minus sign from dt' and another one from the reversal of the integration boundaries

$$= 2\frac{\sin\omega t}{\omega}\Big|_{\Delta-\frac{T}{2}}^{\Delta+\frac{T}{2}} = 2\frac{\sin\omega\left(\Delta+\frac{T}{2}\right) - \sin\omega\left(\Delta-\frac{T}{2}\right)}{\omega}$$

$$= \frac{4}{\omega}\cos\omega\Delta\sin\omega\frac{T}{2}.$$

With (2.29) we finally get:

$$F(\omega) = 2\mathrm{i}\left[\frac{\sin(\omega+\omega_0)\frac{T}{2}\cos(\omega+\omega_0)\Delta}{\omega+\omega_0} - \frac{\sin(\omega-\omega_0)\frac{T}{2}\cos(\omega-\omega_0)\Delta}{\omega-\omega_0}\right]$$

$$= 2\mathrm{i}\left[\frac{\cos(\omega+\omega_0)\Delta\left(\sin\omega\frac{T}{2}\overset{=1}{\cos\omega_0\frac{T}{2}} + \cos\omega\frac{T}{2}\overset{=0}{\sin\omega_0\frac{T}{2}}\right)}{\omega+\omega_0}\right.$$

$$\left. - \frac{\cos(\omega-\omega_0)\Delta\left(\sin\omega\frac{T}{2}\overset{=1}{\cos\omega_0\frac{T}{2}} - \cos\omega\frac{T}{2}\overset{=0}{\sin\omega_0\frac{T}{2}}\right)}{\omega-\omega_0}\right]$$

$$= 2\mathrm{i}\sin\omega\frac{T}{2}\left[\frac{\cos(\omega+\omega_0)\Delta}{\omega+\omega_0} - \frac{\cos(\omega-\omega_0)\Delta}{\omega-\omega_0}\right]$$

$$= \frac{2\mathrm{i}\sin\omega\frac{T}{2}}{\omega^2-\omega_0^2}((\omega-\omega_0)\cos(\omega+\omega_0)\Delta - (\omega+\omega_0)\cos(\omega-\omega_0)\Delta).$$

In order to find the extremes it suffices to calculate:

$$\frac{\mathrm{d}}{\mathrm{d}\Delta}((\omega-\omega_0)\cos(\omega+\omega_0)\Delta - (\omega+\omega_0)\cos(\omega-\omega_0)\Delta) = 0$$

$$(\omega-\omega_0)(-1)(\omega+\omega_0)\sin(\omega+\omega_0)\Delta - (\omega+\omega_0)(\omega-\omega_0)\sin(\omega-\omega_0)\Delta = 0$$

$$\text{or } (\omega^2-\omega_0^2)(\sin(\omega+\omega_0)\Delta - \sin(\omega-\omega_0)\Delta) = 0$$

$$\text{or } (\omega^2-\omega_0^2)\cos\omega\Delta\sin\omega_0\Delta = 0.$$

This is fulfilled for all frequencies ω if $\sin\omega_0\Delta = 0$ or $\omega_0\Delta = l\pi$. With this choice we get finally:

$$F(\omega) = \frac{2\mathrm{i}\sin\omega\frac{T}{2}}{\omega^2-\omega_0^2}\left[(\omega-\omega_0)\left(\cos\omega\Delta\cos\omega_0\Delta - \sin\omega\Delta\overset{=0}{\sin\omega_0\Delta}\right)\right.$$

$$\left. -(\omega+\omega_0)\left(\cos\omega\Delta\cos\omega_0\Delta + \sin\omega\Delta\overset{=0}{\sin\omega_0\Delta}\right)\right]$$

$$= (-\mathrm{i})\frac{1}{2} \int\limits_{-T/2}^{T/2} (\cos(\omega_0 - \omega)t - \cos(\omega_0 + \omega)t)\,\mathrm{d}t$$

$$= (-\mathrm{i}) \int\limits_{0}^{T/2} (\cos(\omega_0 - \omega)t - \cos(\omega_0 + \omega)t)\,\mathrm{d}t$$

$$= (-\mathrm{i})\left(\frac{\sin(\omega_0 - \omega)\frac{T}{2}}{\omega_0 - \omega} - \frac{\sin(\omega_0 + \omega)\frac{T}{2}}{\omega_0 + \omega} \right)$$

$$= (-\mathrm{i})\left(\frac{\sin\omega_0\frac{T}{2} \overset{=0}{\cos\omega\frac{T}{2}} - \overset{=1}{\cos\omega_0\frac{T}{2}} \sin\omega\frac{T}{2}}{\omega_0 - \omega} \right.$$

$$\left. - \frac{\sin\omega_0\frac{T}{2} \overset{=0}{\cos\omega\frac{T}{2}} + \overset{=1}{\cos\omega_0\frac{T}{2}} \sin\omega\frac{T}{2}}{\omega_0 + \omega} \right)$$

$$= \mathrm{i}\sin\omega\frac{T}{2}\left(\frac{1}{\omega_0 - \omega} + \frac{1}{\omega_0 + \omega} \right) = 2\mathrm{i}\sin\frac{\omega T}{2} \times \frac{\omega_0}{\omega_0^2 - \omega^2}.$$

At resonance: $F(\omega_0) = -\mathrm{i}T/2$; $F(-\omega_0) = +\mathrm{i}T/2$; $|F(\pm\omega_0)| = T/2$. This is easily seen by going back to the expressions of the type $\frac{\sin x}{x}$.

For two such pulses centered around $\pm\Delta$ we get:

$$F_{\text{shifted}}(\omega) = 2\mathrm{i}\sin\frac{\omega T}{2} \times \frac{\omega_0}{\omega_0^2 - \omega^2}\left(\mathrm{e}^{\mathrm{i}\omega\Delta} + \mathrm{e}^{-\mathrm{i}\omega\Delta} \right)$$

$$= 4\mathrm{i}\sin\frac{\omega T}{2} \times \frac{\omega_0}{\omega_0^2 - \omega^2}\cos\omega\Delta \quad \longleftarrow \text{"modulation"}.$$

$|F(\omega_0)| = T$ if at resonance: $\omega_0\Delta = l\pi$. In order to maximise $|F(\omega)|$ we require $\omega\Delta = l\pi$; $l = 1, 2, 3, \ldots$; Δ depends on ω!

2.4 Phase-Locked Pulses

This is a textbook case for the Second Shifting Rule! Hence, we start with DC-pulses. This function is even!

$$F_{\text{DC}}(\omega) = \int\limits_{-\Delta-\frac{T}{2}}^{-\Delta+\frac{T}{2}} \cos\omega t\,\mathrm{d}t + \int\limits_{+\Delta-\frac{T}{2}}^{+\Delta+\frac{T}{2}} \cos\omega t\,\mathrm{d}t = 2\int\limits_{\Delta-\frac{T}{2}}^{\Delta+\frac{T}{2}} \cos\omega t\,\mathrm{d}t$$

with $t' = -t$ we get a minus sign from $\mathrm{d}t'$ and another one from the reversal of the integration boundaries

$$= 2\frac{\sin \omega t}{\omega}\Big|_{\Delta-\frac{T}{2}}^{\Delta+\frac{T}{2}} = 2\frac{\sin \omega \left(\Delta + \frac{T}{2}\right) - \sin \omega \left(\Delta - \frac{T}{2}\right)}{\omega}$$

$$= \frac{4}{\omega} \cos \omega \Delta \sin \omega \frac{T}{2}.$$

With (2.29) we finally get:

$$F(\omega) = 2i\left[\frac{\sin(\omega + \omega_0)\frac{T}{2} \cos(\omega + \omega_0)\Delta}{\omega + \omega_0} - \frac{\sin(\omega - \omega_0)\frac{T}{2} \cos(\omega - \omega_0)\Delta}{\omega - \omega_0}\right]$$

$$= 2i\left[\frac{\cos(\omega + \omega_0)\Delta\left(\sin \omega \frac{T}{2} \overset{=1}{\cos \omega_0 \frac{T}{2}} + \cos \omega \frac{T}{2} \overset{=0}{\sin \omega_0 \frac{T}{2}}\right)}{\omega + \omega_0}\right.$$

$$\left. - \frac{\cos(\omega - \omega_0)\Delta\left(\sin \omega \frac{T}{2} \overset{=1}{\cos \omega_0 \frac{T}{2}} - \cos \omega \frac{T}{2} \overset{=0}{\sin \omega_0 \frac{T}{2}}\right)}{\omega - \omega_0}\right]$$

$$= 2i \sin \omega \frac{T}{2}\left[\frac{\cos(\omega + \omega_0)\Delta}{\omega + \omega_0} - \frac{\cos(\omega - \omega_0)\Delta}{\omega - \omega_0}\right]$$

$$= \frac{2i \sin \omega \frac{T}{2}}{\omega^2 - \omega_0^2}((\omega - \omega_0) \cos(\omega + \omega_0)\Delta - (\omega + \omega_0) \cos(\omega - \omega_0)\Delta).$$

In order to find the extremes it suffices to calculate:

$$\frac{d}{d\Delta}((\omega - \omega_0) \cos(\omega + \omega_0)\Delta - (\omega + \omega_0) \cos(\omega - \omega_0)\Delta) = 0$$

$$(\omega - \omega_0)(-1)(\omega + \omega_0) \sin(\omega + \omega_0)\Delta - (\omega + \omega_0)(\omega - \omega_0) \sin(\omega - \omega_0)\Delta = 0$$

$$\text{or } (\omega^2 - \omega_0^2)(\sin(\omega + \omega_0)\Delta - \sin(\omega - \omega_0)\Delta) = 0$$

$$\text{or } (\omega^2 - \omega_0^2) \cos \omega \Delta \sin \omega_0 \Delta = 0.$$

This is fulfilled for all frequencies ω if $\sin \omega_0 \Delta = 0$ or $\omega_0 \Delta = l\pi$. With this choice we get finally:

$$F(\omega) = \frac{2i \sin \omega \frac{T}{2}}{\omega^2 - \omega_0^2}\left[(\omega - \omega_0)\left(\cos \omega \Delta \cos \omega_0 \Delta - \sin \omega \Delta \overset{=0}{\sin \omega_0 \Delta}\right)\right.$$

$$\left. - (\omega + \omega_0)\left(\cos \omega \Delta \cos \omega_0 \Delta + \sin \omega \Delta \overset{=0}{\sin \omega_0 \Delta}\right)\right]$$

$$= \frac{2\mathrm{i} \sin \omega \frac{T}{2}}{\omega^2 - \omega_0^2} (-1)^l \cos \omega \Delta \times 2\omega_0$$

$$= 4\mathrm{i}\omega_0 (-1)^l \frac{\sin \omega \frac{T}{2} \cos \omega \Delta}{\omega^2 - \omega_0^2}.$$

At resonance $\omega = \omega_0$ we get:

$$|F(\omega)| = 4\omega_0 \lim_{\omega \to \omega_0} \frac{\sin \omega \frac{T}{2}}{\omega^2 - \omega_0^2} \qquad \text{with } T = \frac{4\pi}{\omega_0}$$

$$= 4\omega_0 \lim_{\omega \to \omega_0} \frac{\sin 2\pi \frac{\omega}{\omega_0}}{\omega_0^2 \left(\frac{\omega^2}{\omega_0^2} - 1\right)} \qquad \text{with } \alpha = \frac{\omega}{\omega_0}$$

$$= \frac{4}{\omega_0} \lim_{\alpha \to 1} \frac{\sin 2\pi\alpha}{(\alpha - 1)(\alpha + 1)} \qquad \text{with } \beta = \alpha - 1$$

$$= \frac{2}{\omega_0} \lim_{\beta \to 0} \frac{\sin 2\pi(\beta + 1)}{\beta} = \frac{2}{\omega_0} \lim_{\beta \to 0} \left(\frac{\sin 2\pi\beta \overset{=1}{\cos 2\pi} + \cos 2\pi\beta \overset{=0}{\sin 2\pi}}{\beta} \right)$$

$$= \frac{2}{\omega_0} \lim_{\beta \to 0} \frac{2\pi \cos 2\pi\beta}{1} = \frac{4\pi}{\omega_0} = T.$$

For the calculation of the FWHM we better go back to DC-pulses!
   For two pulses separated by 2Δ we get:

$$F_{\mathrm{DC}}(0) = 4\frac{T}{2} \lim_{\omega \to 0} \frac{\sin \omega \frac{T}{2}}{\omega \frac{T}{2}} = 2T$$

$$\text{and } |F_{\mathrm{DC}}(0)|^2 = 4T^2.$$

From $\left(\frac{4}{\omega} \cos \omega \Delta \sin \omega \frac{T}{2}\right)^2 = \frac{1}{2}|F_{\mathrm{DC}}(0)|^2 = 2T^2$ we get (using $\frac{\Delta}{T} = \frac{l}{4}$):

$$16 \cos^2 \frac{\omega T l}{4} \sin^2 \frac{\omega T}{2} = 2T^2 \omega^2 \qquad \text{with } x = \frac{\omega T}{4}$$

$$\cos^2 x l \sin^2 2x = 2x^2.$$

For $l = 1$ we get:

$$\cos^2 x \sin^2 2x = 2x^2$$

$$\text{or } \cos x \sin 2x = \sqrt{2}x$$

$$\cos x \times 2 \sin x \cos x = \sqrt{2}x$$

$$\cos^2 x \sin x = \frac{x}{\sqrt{2}}.$$

The solution of this transcendental equation yields:

$$\Delta\omega = \frac{4.265}{T} \quad \text{with } \Delta = \frac{T}{4}.$$

For $l = 2$ we get:

$$\cos^2 2x \sin^2 2x = 2x^2$$
$$\text{or } \cos 2x \sin 2x = \sqrt{2}x$$
$$\frac{1}{2}\sin 4x = \sqrt{2}x$$
$$\sin 4x = 2\sqrt{2}x.$$

The solution of this transcendental equation yields:

$$\Delta\omega = \frac{2.783}{T} \quad \text{with } \Delta = \frac{T}{2}.$$

These values for the FWHM should be compared with the value for a single DC-pulse (see (3.12)):

$$\Delta\omega = \frac{5.566}{T}.$$

The Fourier transform of such a double pulse represents the frequency spectrum which is available for excitation in a resonant absorption experiment. In radiofrequency spectroscopy this is called the Ramsey technique, medical doctors would call it fractionated medication.

2.5 Tricky Convolution

We want to calculate $h(t) = f_1(t) \otimes f_2(t)$. Let's do it the other way round. We know from the Convolution Theorem that the Fourier transform of the convolution integral is merely a product of the individual Fourier transforms, i.e.

$$f_{1,2}(t) = \frac{\sigma_{1,2}}{\pi}\frac{1}{\sigma_{1,2}^2 + t^2} \quad \leftrightarrow \quad F_{1,2}(\omega) = e^{-\sigma_{1,2}|\omega|}.$$

Check:

$$F(\omega) = \frac{2\sigma}{\pi}\int_0^\infty \frac{\cos\omega t}{\sigma^2 + t^2}dt$$

$$= \frac{2}{\pi\sigma}\int_0^\infty \frac{\cos\omega t}{1 + (t/\sigma)^2}dt$$

$$= \frac{2}{\pi\sigma} \int\limits_{0}^{\infty} \frac{\cos(\omega\sigma t')}{1 + t'^2} \sigma dt' \quad \text{with } t' = \frac{t}{\sigma}$$

$$= \frac{2}{\pi} \frac{\pi}{2} e^{-\sigma|\omega|} = e^{-\sigma|\omega|}.$$

No wonder, it's just the inverse problem of (2.18).

Hence, $H(\omega) = \exp(-\sigma_1|\omega|)\exp(-\sigma_2|\omega|) = \exp(-(\sigma_1 + \sigma_2)|\omega|)$. The inverse transformation yields:

$$h(t) = \frac{2}{2\pi} \int\limits_{0}^{\infty} e^{-(\sigma_1+\sigma_2)\omega} \cos\omega t d\omega$$

$$= \frac{1}{\pi} \frac{\sigma_1 + \sigma_2}{(\sigma_1 + \sigma_2)^2 + t^2},$$

i.e. another Lorentzian with $\sigma_{\text{total}} = \sigma_1 + \sigma_2$.

2.6 Even Trickier

We have:

$$f_1(t) = \frac{1}{\sigma_1\sqrt{2\pi}} e^{-\frac{1}{2}\frac{t^2}{\sigma_1^2}} \quad \leftrightarrow \quad F_1(\omega) = e^{-\frac{1}{2}\sigma_1^2\omega^2}$$

and:

$$f_2(t) = \frac{1}{\sigma_2\sqrt{2\pi}} e^{-\frac{1}{2}\frac{t^2}{\sigma_2^2}} \quad \leftrightarrow \quad F_2(\omega) = e^{-\frac{1}{2}\sigma_2^2\omega^2}.$$

We want to calculate $h(t) = f_1(t) \otimes f_2(t)$.

We have $H(\omega) = \exp\left(\frac{1}{2}\left(\sigma_1^2 + \sigma_2^2\right)\omega^2\right)$. This we have to backtransform in order to get the convolution integral:

$$h(t) = \frac{1}{2\pi} \int\limits_{-\infty}^{+\infty} e^{-\frac{1}{2}(\sigma_1^2+\sigma_2^2)\omega^2} e^{+i\omega t} d\omega$$

$$= \frac{1}{\pi} \int\limits_{0}^{\infty} e^{-\frac{1}{2}(\sigma_1^2+\sigma_2^2)\omega^2} \cos\omega t d\omega$$

$$= \frac{1}{\pi} \frac{\sqrt{\pi}}{2\frac{1}{\sqrt{2}}\sqrt{\sigma_1^2 + \sigma_2^2}} e^{-\frac{t^2}{4\frac{1}{2}(\sigma_1^2+\sigma_2^2)}}$$

$$= \frac{1}{\sqrt{2\pi}} \frac{1}{\sqrt{\sigma_1^2 + \sigma_2^2}} e^{-\frac{1}{2}\frac{t^2}{\sigma_1^2+\sigma_2^2}}$$

$$= \frac{1}{\sqrt{2\pi}\,\sigma_{\text{total}}} \mathrm{e}^{-\frac{1}{2}\frac{t^2}{\sigma_{\text{total}}^2}} \quad \text{with } \sigma_{\text{total}}^2 = \sigma_1^2 + \sigma_2^2.$$

Hence, it is again a Gaussian with the σ's squared added. The calculation of the convolution integral directly is much more tedious:

$$f_1(t) \otimes f_2(t) = \frac{1}{\sigma_1 \sigma_2 2\pi} \int\limits_{-\infty}^{+\infty} \mathrm{e}^{-\frac{1}{2}\frac{\xi^2}{\sigma_1^2}} \mathrm{e}^{-\frac{1}{2}\frac{(t-\xi)^2}{\sigma_2^2}} \, \mathrm{d}\xi$$

with the exponent:

$$-\frac{1}{2}\left[\frac{\xi^2}{\sigma_1^2} + \frac{\xi^2}{\sigma_2^2} - \frac{2t\xi}{\sigma_2^2} + \frac{t^2}{\sigma_2^2} \right]$$

$$= -\frac{1}{2}\left[\left(\frac{1}{\sigma_1^2} + \frac{1}{\sigma_2^2} \right)\left(\xi^2 - \frac{2t\xi}{\sigma_2^2}\frac{1}{\frac{1}{\sigma_1^2} + \frac{1}{\sigma_2^2}} \right) + \frac{t^2}{\sigma_2^2} \right]$$

$$= -\frac{1}{2}\left[\left(\frac{1}{\sigma_1^2} + \frac{1}{\sigma_2^2} \right)\left(\xi^2 - \frac{2t\xi\sigma_1^2}{\sigma_1^2 + \sigma_2^2} + \frac{t^2\sigma_1^4}{(\sigma_1^2 + \sigma_2^2)^2} - \frac{t^2\sigma_1^4}{(\sigma_1^2 + \sigma_2^2)^2} \right) + \frac{t^2}{\sigma_2^2} \right]$$

$$= -\frac{1}{2}\left[\left(\frac{1}{\sigma_1^2} + \frac{1}{\sigma_2^2} \right)\left(\xi - \frac{t\sigma_1^2}{\sigma_1^2 + \sigma_2^2} \right)^2 - \frac{(\sigma_1^2 + \sigma_2^2)}{\sigma_1^2 \sigma_2^2}\frac{t^2\sigma_1^4}{(\sigma_1^2 + \sigma_2^2)^2} + \frac{t^2}{\sigma_2^2} \right]$$

$$= -\frac{1}{2}\left[\left(\frac{1}{\sigma_1^2} + \frac{1}{\sigma_2^2} \right)\left(\xi - \frac{t\sigma_1^2}{\sigma_1^2 + \sigma_2^2} \right)^2 - \frac{t^2\sigma_1^2}{\sigma_2^2(\sigma_1^2 + \sigma_2^2)} + \frac{t^2}{\sigma_2^2} \right]$$

$$= -\frac{1}{2}\left[\left(\frac{1}{\sigma_1^2} + \frac{1}{\sigma_2^2} \right)\left(\xi - \frac{t\sigma_1^2}{\sigma_1^2 + \sigma_2^2} \right)^2 + \frac{t^2}{\sigma_2^2}\left(1 - \frac{\sigma_1^2}{\sigma_1^2 + \sigma_2^2} \right) \right]$$

$$= -\frac{1}{2}\left[\left(\frac{1}{\sigma_1^2} + \frac{1}{\sigma_2^2} \right)\left(\xi - \frac{t\sigma_1^2}{\sigma_1^2 + \sigma_2^2} \right)^2 + \frac{t^2}{\sigma_2^2}\frac{\sigma_2^2}{\sigma_1^2 + \sigma_2^2} \right]$$

$$= -\frac{1}{2}\left[\left(\frac{1}{\sigma_1^2} + \frac{1}{\sigma_2^2} \right)\left(\xi - \frac{t\sigma_1^2}{\sigma_1^2 + \sigma_2^2} \right)^2 + \frac{t^2}{\sigma_1^2 + \sigma_2^2} \right]$$

hence:

$$f_1(t) \otimes f_2(t) = \frac{1}{\sigma_1 \sigma_2 2\pi} \mathrm{e}^{-\frac{1}{2}\frac{t^2}{\sigma_1^2 + \sigma_2^2}} \int\limits_{-\infty}^{+\infty} \mathrm{e}^{-\frac{1}{2}\left(\frac{1}{\sigma_1^2} + \frac{1}{\sigma_2^2} \right)\left(\xi - \frac{t\sigma_1^2}{\sigma_1^2 + \sigma_2^2} \right)^2} \, \mathrm{d}\xi$$

$$\text{with } \xi - \frac{t\sigma_1^2}{\sigma_1^2 + \sigma_2^2} = \xi'$$

$$= \frac{1}{\sigma_1\sigma_2 2\pi} e^{-\frac{1}{2}\frac{t^2}{\sigma_1^2+\sigma_2^2}} \int\limits_{-\infty}^{+\infty} e^{-\frac{1}{2}\left(\frac{1}{\sigma_1^2}+\frac{1}{\sigma_2^2}\right)\xi'^2} d\xi'$$

$$= \frac{1}{\sigma_1\sigma_2 2\pi} e^{-\frac{1}{2}\frac{t^2}{\sigma_1^2+\sigma_2^2}} \frac{\sqrt{\pi}}{2} \frac{2}{\frac{1}{\sqrt{2}}\sqrt{\frac{1}{\sigma_1^2}+\frac{1}{\sigma_2^2}}}$$

$$= \frac{1}{\sqrt{2\pi}} e^{-\frac{1}{2}\frac{t^2}{\sigma_1^2+\sigma_2^2}} \frac{1}{\sigma_1\sigma_2} \frac{\sigma_1\sigma_2}{\sqrt{\sigma_1^2+\sigma_2^2}}$$

$$= \frac{1}{\sqrt{2\pi}} \frac{1}{\sigma_{\text{total}}} e^{-\frac{1}{2}\frac{t^2}{\sigma_{\text{total}}^2}} \quad \text{with } \sigma_{\text{total}}^2 = \sigma_1^2 + \sigma_2^2.$$

2.7 Voigt Profile (for Gourmets only)

$$f_1(t) = \frac{\sigma_1}{\pi} \frac{1}{\sigma_1^2+t^2} \qquad \leftrightarrow \quad F_1(\omega) = e^{-\sigma_1|\omega|}$$

$$f_2(t) = \frac{1}{\sigma_2\sqrt{2\pi}} e^{-\frac{1}{2}\frac{t^2}{\sigma_2^2}} \quad \leftrightarrow \quad F_2(\omega) = e^{-\frac{1}{2}\sigma_2^2\omega^2}$$

$$H(\omega) = e^{-\sigma_1|\omega|} e^{-\frac{1}{2}\sigma_2^2\omega^2}.$$

The inverse transformation is a nightmare! Note that $H(\omega)$ is an even function.

$$h(t) = \frac{1}{2\pi} 2 \int\limits_0^\infty e^{-\sigma_1\omega} e^{-\frac{1}{2}\sigma_2^2\omega^2} \cos\omega t \, d\omega$$

$$= \frac{1}{\pi} \frac{1}{2\left(2\frac{1}{2}\sigma_2^2\right)^{\frac{1}{2}}} \exp\left(\frac{\sigma_1^2 - t^2}{8\frac{1}{2}\sigma_2^2}\right)$$

$$\times \Gamma(1) \left\{ \exp\left(-\frac{i\sigma_1 t}{4\frac{1}{2}\sigma_2^2}\right) D_{-1}\left(\frac{\sigma_1 - it}{\sqrt{2\frac{1}{2}\sigma_2^2}}\right) \right.$$

$$\left. + \exp\left(\frac{i\sigma_1 t}{4\frac{1}{2}\sigma_2^2}\right) D_{-1}\left(\frac{\sigma_1 + it}{\sqrt{2\frac{1}{2}\sigma_2^2}}\right) \right\}$$

$$= \frac{1}{2\pi}\frac{1}{\sigma_2}\exp\left(\frac{\sigma_1^2 - t^2}{4\sigma_2^2}\right)\left\{\exp\left(-\frac{i\sigma_1 t}{2\sigma_2^2}\right)D_{-1}\left(\frac{\sigma_1 - it}{\sigma_2}\right) + \text{c.c.}\right\}$$

with $D_{-1}(z)$ denoting a parabolic cylinder function. The complex conjugate ("c.c.") ensures that $h(t)$ is real. A similar situation shows up in (3.32) where we truncate a Gaussian. Here, we have a cusp in $H(\omega)$. What a messy lineshape for a Lorentzian spectral line and a spectrometer with a Gaussian resolution function!

Among spectroscopists, this lineshape is known as the "Voigt profile". The parabolic cylinder function $D_{-1}(z)$ can be expressed in terms of the complementary error function:

$$D_{-1}(z) = e^{\frac{z^2}{4}}\sqrt{\frac{\pi}{2}}\text{erfc}\left(\frac{z}{\sqrt{2}}\right).$$

Hence, we can write:

$$
\begin{aligned}
h(t) &= \frac{1}{2\pi\sigma_2}\sqrt{\frac{\pi}{2}}e^{\left(\frac{\sigma_1 - it}{\sigma_2}\right)^2\frac{1}{4}}\text{erfc}\left(\frac{\sigma_1 - it}{\sqrt{2}\sigma_2}\right)e^{+\frac{\sigma_1^2 - t^2}{4\sigma_2^2}}e^{-\frac{i\sigma_1 t}{2\sigma_2^2}} \\
&\quad + \frac{1}{2\pi\sigma_2}\sqrt{\frac{\pi}{2}}e^{\left(\frac{\sigma_1 + it}{\sigma_2}\right)^2\frac{1}{4}}\text{erfc}\left(\frac{\sigma_1 + it}{\sqrt{2}\sigma_2}\right)e^{+\frac{\sigma_1^2 - t^2}{4\sigma_2^2}}e^{+\frac{i\sigma_1 t}{2\sigma_2^2}} \\
&= \frac{1}{\sqrt{2\pi}2\sigma_2}\left\{e^{\frac{1}{4\sigma_2^2}[\sigma_1^2 - 2it\sigma_1 - t^2 + \sigma_1^2 - t^2 - 2i\sigma_1 t]}\text{erfc}\left(\frac{\sigma_1 - it}{\sqrt{2}\sigma_2}\right)\right. \\
&\quad \left. + e^{\frac{1}{4\sigma_2^2}[\sigma_1^2 + 2it\sigma_1 - t^2 + \sigma_1^2 - t^2 + 2i\sigma_1 t]}\text{erfc}\left(\frac{\sigma_1 + it}{\sqrt{2}\sigma_2}\right)\right\} \\
&= \frac{1}{\sqrt{2\pi}2\sigma_2}\left\{e^{\frac{1}{2\sigma_2^2}(\sigma_1^2 - 2it\sigma_1 - t^2)}\text{erfc}\left(\frac{\sigma_1 - it}{\sqrt{2}\sigma_2}\right)\right. \\
&\quad \left. + e^{\frac{1}{2\sigma_2^2}(\sigma_1^2 + 2it\sigma_1 - t^2)}\text{erfc}\left(\frac{\sigma_1 + it}{\sqrt{2}\sigma_2}\right)\right\} \\
&= \frac{1}{\sqrt{2\pi}2\sigma_2}\left\{e^{\left(\frac{\sigma_1 - it}{\sqrt{2}\sigma_2}\right)^2}\text{erfc}\left(\frac{\sigma_1 - it}{\sqrt{2}\sigma_2}\right) + e^{\left(\frac{\sigma_1 + it}{\sqrt{2}\sigma_2}\right)^2}\text{erfc}\left(\frac{\sigma_1 + it}{\sqrt{2}\sigma_2}\right)\right\} \\
&= \frac{1}{\sqrt{2\pi}2\sigma_2}\text{erfc}\left(\frac{\sigma_1 - it}{\sqrt{2}\sigma_2}\right)e^{\left(\frac{\sigma_1 - it}{\sqrt{2}\sigma_2}\right)^2} + \text{c.c.}
\end{aligned}
$$

2.8 Derivable

The function is mixed. We know that $\frac{dF(\omega)}{d\omega} = -i\text{FT}(tf(t))$ with $f(t) = e^{-t/\tau}$ for $t \geq 0$ (see 2.58), and we know its Fourier transform (see 2.21) $F(\omega) = 1/(\lambda + i\omega)$.

Hence:

$$G(\omega) = i\frac{d}{d\omega}\left(\frac{1}{\lambda + i\omega}\right)$$

$$= i\frac{(-i)}{(\lambda + i\omega)^2} = \frac{1}{(\lambda + i\omega)^2}$$

$$= \frac{(\lambda - i\omega)^2}{(\lambda + i\omega)^2(\lambda - i\omega)^2} = \frac{\lambda^2 - 2i\omega\lambda - \omega^2}{(\lambda^2 + \omega^2)^2}$$

$$= \frac{\lambda^2 - \omega^2}{(\lambda^2 + \omega^2)^2} - \frac{2i\omega\lambda}{(\lambda^2 + \omega^2)^2}$$

$$= \frac{(\lambda^2 - \omega^2) - 2i\omega\lambda}{(\lambda^2 + \omega^2)^2}.$$

Inverse transformation:

$$g(t) = \frac{1}{2\pi}\int_{-\infty}^{\infty}\frac{e^{i\omega t}}{(\lambda + i\omega)^2}d\omega$$

Real part: $$\frac{1}{2\pi}2\int_0^{\infty}\cos\omega t\frac{\lambda^2 - \omega^2}{(\lambda^2 + \omega^2)^2}d\omega$$

Imaginary part: $$\frac{1}{2\pi}2\int_0^{\infty}\sin\omega t\frac{(-2)\omega\lambda}{(\lambda^2 + \omega^2)^2}d\omega; \qquad (\omega\sin\omega t \text{ is even in } \omega!).$$

Hint: Reference [8, Nos 3.769.1, 3.769.2] $\nu = 2$; $\beta = \lambda$; $x = \omega$:

$$\frac{1}{(\lambda + i\omega)^2} + \frac{1}{(\lambda - i\omega)^2} = \frac{2(\lambda^2 - \omega^2)}{(\lambda^2 + \omega^2)^2}$$

$$\frac{1}{(\lambda + i\omega)^2} - \frac{1}{(\lambda - i\omega)^2} = \frac{-4i\omega\lambda}{(\lambda^2 + \omega^2)^2}$$

$$\int_0^{\infty}\frac{(\lambda^2 - \omega^2)}{(\lambda^2 + \omega^2)^2}\cos\omega t d\omega = \frac{\pi}{2}te^{-\lambda t}$$

$$\int_0^{\infty}\frac{-2i\omega\lambda}{(\lambda^2 + \omega^2)^2}\sin\omega t d\omega = \frac{\pi}{2}ite^{-\lambda t}$$

from real part   from imaginary part
$$\frac{1}{\pi}\frac{\pi}{2}te^{-\lambda t} \quad + \qquad \frac{1}{\pi}\frac{\pi}{2}te^{-\lambda t} \quad = te^{-\lambda t} \qquad \text{for } t > 0.$$

2.9 Nothing Gets Lost

First, we note that the integral is an even function and we can write:

$$\int_0^\infty \frac{\sin^2 a\omega}{\omega^2} d\omega = \frac{1}{2} \int_{-\infty}^{+\infty} \frac{\sin^2 a\omega}{\omega^2} d\omega.$$

Next, we identify $\sin a\omega/\omega$ with $F(\omega)$, the Fourier transform of the "rectangular function" with $a = T/2$ (and a factor of 2 smaller).

The inverse transform yields:

$$f(t) = \begin{cases} 1/2 \text{ for } -a \le t \le a \\ 0 \quad \text{else} \end{cases}$$

$$\text{and} \int_{-a}^{+a} |f(t)|^2 dt = \frac{1}{4} 2a = \frac{a}{2}.$$

Finally, Parseval's theorem gives:

$$\frac{a}{2} = \frac{1}{2\pi} \int_{-\infty}^{+\infty} \frac{\sin^2 a\omega}{\omega^2} d\omega$$

$$\text{or} \int_{-\infty}^{\infty} \frac{\sin^2 a\omega}{\omega^2} d\omega = \frac{2\pi a}{2} = \pi a$$

$$\text{or} \int_0^{\infty} \frac{\sin^2 a\omega}{\omega^2} d\omega = \frac{\pi a}{2}.$$

Playground of Chap. 3

3.1 Squared

$f(\omega) = T \sin(\omega T/2)/(\omega T/2)$. At $\omega = 0$ we have $F(0) = T$. This function drops to $T/2$ at a frequency ω defined by the following transcendental equation:

$$\frac{T}{2} = T \frac{\sin(\omega T/2)}{\omega T/2}$$

with $x = \omega T/2$ we have $x/2 = \sin x$ with the solution $x = 1.8955$, hence $\omega_{3dB} = 3.791/T$. With a pocket calculator we might have done the following:

x	$\sin x$	$x/2$
1.5	0.997	0.75
1.4	0.985	0.7
1.6	0.9995	0.8
1.8	0.9738	0.9
1.85	0.9613	0.925
1.88	0.9526	0.94
1.89	0.9495	0.945
1.895	0.9479	0.9475
1.896	0.9476	0.948
1.8955	0.94775	0.94775

The total width is $\Delta\omega = 7.582/T$.

For $F^2(\omega)$ we had $\Delta\omega = 5.566/T$; hence the 3 dB-bandwidth of $F(\omega)$ is a factor of 1.362 larger than that of $F^2(\omega)$, about 4 % less than $\sqrt{2} = 1.414$.

3.2 Let's Gibbs Again

There are tiny steps at the interval boundaries, hence we have -6 dB/octave.

3.3 Expander

Blackman–Harris window:

$$f(t) = \begin{cases} \displaystyle\sum_{n=0}^{3} a_n \cos \frac{2\pi n t}{T} & \text{for } -T/2 \le t \le T/2 \\ \\ 0 & \text{else} \end{cases}.$$

From the expansion of the cosines we get (in the interval $-T/2 \le t \le T/2$):

$$f(t) = \sum_{n=0}^{3} a_n \left(1 - \frac{1}{2!}\left(\frac{2\pi n t}{T}\right)^2 + \frac{1}{4!}\left(\frac{2\pi n t}{T}\right)^4 - \frac{1}{6!}\left(\frac{2\pi n t}{T}\right)^6 + \dots \right)$$

$$= \sum_{k=0}^{\infty} b_k \left(\frac{t}{T/2}\right)^{2k}.$$

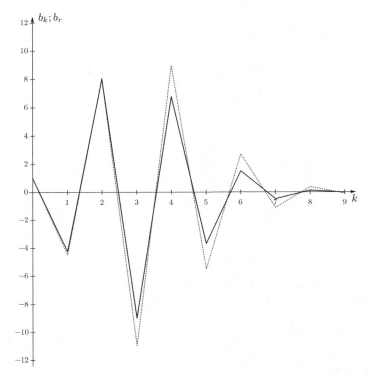

Fig. A.7 Expansion coefficients b_k for the Blackman–Harris window (-74 dB) *(dotted line)* and expansion coefficients b_r for the Kaiser–Bessel window ($\beta = 9$) *(solid line)*. There are even powers of t only, i.e. the coefficient b_6 corresponds to t^{12}

Inserting the coefficients a_n for the -74 dB-window including the second option of the corrections on p. 86 we get:

k	b_k
0	$+1.0000$
1	-4.4888
2	$+8.0592$
3	-10.9471
4	$+8.9791$
5	-5.4918
6	$+2.7106$
7	-1.1105
8	$+0.3761$
9	-0.1047

The coefficients are displayed in Fig. A.7. Note that at the interval boundaries $t = \pm T/2$ we should have $\sum_{k=0}^{\infty} b_k = 0$. The first ten terms add up to -0.0196.

Next, we calculate:

$$I_0(z) = \sum_{k=0}^{\infty} \frac{\left(\frac{1}{4}z^2\right)^k}{(k!)^2}$$

for $z = 9$.

k	$(4.5^k/k!)^2$
0	1.000
1	20.250
2	102.516
3	230.660
4	291.929
5	236.463
6	133.010
7	54.969
8	17.392
9	4.348

Summing up the first ten terms, we get 1,092.5, close to the exact value of 1,093.588.

Next, we have to expand the numerator of the Kaiser–Bessel window function.

$$I(9)f(t) = \sum_{k=0}^{\infty} \frac{\left[\frac{81}{4}\left(1 - \left(\frac{2t}{T}\right)^2\right)\right]^k}{(k!)^2}$$

$$= \sum_{k=0}^{\infty} \frac{\left(\frac{81}{4}\right)^k}{(k!)^2}\left(1 - \left(\frac{2t}{T}\right)^2\right)^k \quad \text{with } \left(\frac{2t}{T}\right)^2 = y$$

$$= \sum_{k=0}^{\infty} \left[\frac{\left(\frac{9}{2}\right)^k}{k!}\right]^2 (1-y)^k$$

$$\left[\text{with binomial formula } (1-y)^k = \sum_{r=0}^{k}\binom{k}{r}(-1)^r y^r = \sum_{r=0}^{k}\frac{k!}{r!(k-r)!}(-y)^r\right]$$

$$= \sum_{k=0}^{\infty} \left[\frac{\left(\frac{9}{2}\right)^k}{k!}\right]^2 \sum_{r=0}^{k}\frac{k!}{r!(k-r)!}(-y)^r$$

$$= \sum_{\substack{k=0 \\ r=0}}^{\infty} \left[\frac{\left(\frac{9}{2}\right)^k}{k!}\right]^2 + \sum_{\substack{k=1 \\ r=1}}^{\infty} \left[\frac{\left(\frac{9}{2}\right)^k}{k!}\right]^2 \overbrace{\frac{k!}{(k-1)!}}^{=k}(-y)^1$$

$$+ \sum_{k=2}^{\infty} \left[\frac{\left(\frac{9}{2}\right)^k}{k!} \right]^2 \overbrace{\frac{k!}{2!(k-2)!}}^{=k(k-1)/2} y^2$$
$$\underbrace{\qquad}_{r=2}$$

$$+ \sum_{k=3}^{\infty} \left[\frac{\left(\frac{9}{2}\right)^k}{k!} \right]^2 \overbrace{\frac{k!}{3!(k-3)!}}^{=k(k-1)(k-2)/6} (-y)^3 + \cdots$$
$$\underbrace{\qquad}_{r=3}$$

$$= \sum_{r=0}^{\infty} b_r \left(\frac{t}{T/2} \right)^{2r}$$

(*Note*: For integer and negative k we have $k! = \pm\infty$ and $0! = 1$.).

Here, the calculation of each expansion coefficient b_r requires (in principle) the calculation of an infinite series. We truncate the series at $k = 9$. For $r = 0$ up to $r = 9$ we get:

r	b_r
0	+1.0000
1	−4.2421
2	+8.0039
3	−8.9811
4	+6.7708
5	−3.6767
6	+1.5063
7	−0.4816
8	+0.1233
9	−0.0258

These coefficients are displayed in Fig. A.7. Note, that at the interval boundaries $t = \pm T/2$ the coefficients b_r do no longer have to add up to 0 exactly. Figure A.7 shows why the Blackman–Harris (-74 dB) window and the Kaiser–Bessel ($\beta = 9$) window have similar properties.

3.4 Minorities

a. For a rectangular window we have $\Delta\omega = 5.566/T = 50$ Mrad/s from which we get $T = 111.32$ ns.
b. The suspected signal is at 600 Mrad/s, i.e. 4 times the FWHM away from the central peak.

The rectangular window is not good for the detection. The triangular window has a factor $8.016/5.566 = 1.44$ larger FWHM, i.e. our suspected peak is 2.78 times the FWHM away from the central peak. A glance to Fig. 3.2 tells you, that this window is also not good. The cosine window has only a factor of $7.47/5.566 = 1.34$ larger FWHM, but is still not good enough. For the $\cos^2$-window we have a factor of $9.06/5.566 = 1.63$ larger FWHM, i.e. only 2.45 times the FWHM away from the central peak. This means, that -50 dB, 2.45 times the FWHM higher than the central peak, is still not detectable with this window. Similarly, the Hamming window is not good enough. The Gauss window as described in Sect. 3.7 would be a choice because $\Delta\omega\pi \sim 9.06$, but the sidelobe suppression just suffices.

The Kaiser–Bessel window with $\beta = 8$ has $\Delta\omega T \sim 10$, but sufficient sidelobe suppression, and, of course, both Blackman–Harris windows would be adequate.

Playground of Chap. 4

4.1 Correlated

$\{h_k\} = (\text{const.}/N) \sum_{l=0}^{N-1} f_l$, independent of k if $\sum_{l=0}^{N-1} f_l$ vanishes (i.e. the average is 0) then $\{h_k\} = 0$ for all k, otherwise $\{h_k\} = \text{const.} \times \langle f_l \rangle$ for all k (see Fig. A.8).

4.2 No Common Ground

$$h_k = \frac{1}{N} \sum_{l=0}^{N-1} f_l g_{l+k}^*$$

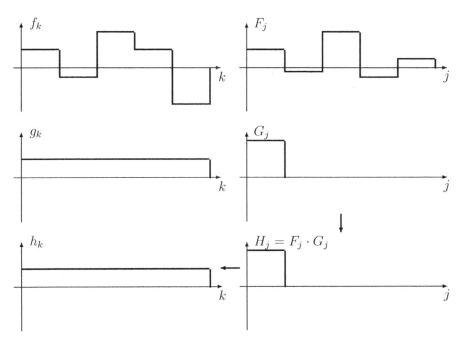

Fig. A.8 An arbitrary f_k (*top left*) and its Fourier transform F_j (*top right*). A constant g_k (*middle left*) and its Fourier transform G_j (*middle right*). The product of $H_j = F_j G_j$ (*bottom right*) and its inverse transform h_k (*bottom left*)

we don't need * here.

$$h_0 = \frac{1}{4}(f_0 g_0 + f_1 g_1 + f_2 g_2 + f_3 g_3)$$
$$= \frac{1}{4}(1 \times 1 + 0 \times (-1) + (-1) \times 1 + 0 \times (-1)) = 0$$
$$h_1 = \frac{1}{4}(f_0 g_1 + f_1 g_2 + f_2 g_3 + f_3 g_0)$$
$$= \frac{1}{4}(1 \times (-1) + 0 \times 1 + (-1) \times (-1) + 0 \times 1) = 0$$
$$h_2 = \frac{1}{4}(f_0 g_2 + f_1 g_3 + f_2 g_0 + f_3 g_1)$$
$$= \frac{1}{4}(1 \times 1 + 0 \times (-1) + (-1) \times 1 + 0 \times (-1)) = 0$$
$$h_3 = \frac{1}{4}(f_0 g_3 + f_1 g_0 + f_2 g_1 + f_3 g_2)$$
$$= \frac{1}{4}(1 \times (-1) + 0 \times 1 + (-1) \times (-1) + 0 \times 1) = 0$$

f corresponds to half the Nyquist frequency and g corresponds to the Nyquist frequency. Their cross correlation vanishes. The FT of $\{f_k\}$ is $\{F_j\} = \{0, 1/2, 0, 1/2\}$, the FT of $\{g_k\}$ is $\{G_j\} = \{0, 0, 1, 0\}$. The multiplication of $F_j G_j$ shows that there is nothing in common:

$$F_j G_j = \{0, 0, 0, 0\} \text{ and, hence, } \{h_k\} = \{0, 0, 0, 0\}.$$

4.3 Brotherly

$$F_0 = \frac{1}{2}$$
$$F_1 = \frac{1}{4}\left(1 + 0 \times e^{-\frac{2\pi i \times 1}{4}} + 1 \times e^{-\frac{2\pi i \times 2}{4}} + 0 \times e^{-\frac{2\pi i \times 3}{4}}\right)$$
$$= \frac{1}{4}(1 + 0 + (-1) + 0) = 0$$
$$F_2 = \frac{1}{4}\left(1 + 0 \times e^{-\frac{2\pi i \times 2}{4}} + 1 \times e^{-\frac{2\pi i \times 4}{4}} + 0 \times e^{-\frac{2\pi i \times 6}{4}}\right)$$
$$= \frac{1}{4}(1 + 0 + 1 + 0) = \frac{1}{2}$$
$$F_3 = 0$$
$$G_j = \{0, 0, 1, 0\} \qquad \text{Nyquist frequency}$$
$$H_j = F_j G_j = \{0, 0, 1/2, 0\}.$$

Inverse transformation:

$$h_k = \sum_{j=0}^{N-1} H_j W_N^{+kj} \qquad W_4^{+kj} = e^{\frac{2\pi i k j}{N}}.$$

Hence:

$$h_k = \sum_{j=0}^{3} H_j e^{\frac{2\pi i k j}{4}} = \sum_{j=0}^{3} H_j e^{i\frac{\pi k j}{2}}$$

$$h_0 = H_0 + H_1 + H_2 + H_3 = \frac{1}{2}$$

$$h_1 = H_0 + H_1 \times i + H_2 \times (-1) + H_3 \times (-i) = -\frac{1}{2}$$

$$h_2 = H_0 + H_1 \times (-1) + H_2 \times 1 + H_3 \times (-1) = \frac{1}{2}$$

$$h_3 = H_0 + H_1 \times (-i) + H_2 \times (-1) + H_3 \times i = -\frac{1}{2}.$$

Figure A.9 is the graphical illustration.

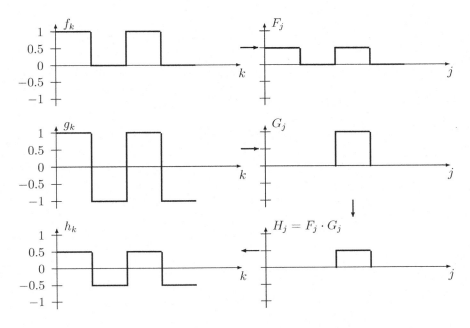

Fig. A.9 Nyquist frequency plus const.$= 1/2$ (*top left*) and its Fourier transform F_j (*top right*). Nyquist frequency (*middle left*) and its Fourier transform G_j (*middle right*). Product of $H_j = F_j G_j$ (*bottom right*) and its inverse transform (*bottom left*)

4.4 Autocorrelated

$N = 6$, real input:

$$h_k = \frac{1}{6} \sum_{l=0}^{5} f_l f_{l+k}$$

$$h_0 = \frac{1}{6} \sum_{l=0}^{5} f_l^2 = \frac{1}{6}(1 + 4 + 9 + 4 + 1) = \frac{19}{6}$$

$$h_1 = \frac{1}{6}(f_0 f_1 + f_1 f_2 + f_2 f_3 + f_3 f_4 + f_4 f_5 + f_5 f_0)$$

$$= \frac{1}{6}(0 \times 1 + 1 \times 2 + 2 \times 3 + 3 \times 2 + 2 \times 1 + 1 \times 0)$$

$$= \frac{1}{6}(2 + 6 + 6 + 2) = \frac{16}{6}$$

$$h_2 = \frac{1}{6}(f_0 f_2 + f_1 f_3 + f_2 f_4 + f_3 f_5 + f_4 f_0 + f_5 f_1)$$

$$= \frac{1}{6}(0 \times 2 + 1 \times 3 + 2 \times 2 + 3 \times 1 + 2 \times 0 + 1 \times 1)$$

$$= \frac{1}{6}(3 + 4 + 3 + 1) = \frac{11}{6}$$

$$h_3 = \frac{1}{6}(f_0 f_3 + f_1 f_4 + f_2 f_5 + f_3 f_0 + f_4 f_1 + f_5 f_2)$$

$$= \frac{1}{6}(0 \times 3 + 1 \times 2 + 2 \times 1 + 3 \times 0 + 2 \times 1 + 1 \times 2)$$

$$= \frac{1}{6}(2 + 2 + 2 + 2) = \frac{8}{6}$$

$$h_4 = \frac{1}{6}(f_0 f_4 + f_1 f_5 + f_2 f_0 + f_3 f_1 + f_4 f_2 + f_5 f_3)$$

$$= \frac{1}{6}(0 \times 2 + 1 \times 1 + 2 \times 0 + 3 \times 1 + 2 \times 2 + 1 \times 3)$$

$$= \frac{1}{6}(1 + 3 + 4 + 3) = \frac{11}{6}$$

$$h_5 = \frac{1}{6}(f_0 f_5 + f_1 f_0 + f_2 f_1 + f_3 f_2 + f_4 f_3 + f_5 f_4)$$

$$= \frac{1}{6}(0 \times 1 + 1 \times 0 + 2 \times 1 + 3 \times 2 + 2 \times 3 + 1 \times 2)$$

$$= \frac{1}{6}(2 + 6 + 6 + 2) = \frac{16}{6}.$$

FT of $\{f_k\}$: $N = 6$, $f_k = f_{-k} = f_{6-k} \rightarrow$ even!

$$F_j = \frac{1}{6} \sum_{k=0}^{5} f_k \cos \frac{2\pi k j}{6} = \frac{1}{6} \sum_{k=0}^{5} f_k \cos \frac{\pi k j}{3}$$

$$F_0 = \frac{1}{6}(0 + 1 + 2 + 3 + 2 + 1) = \frac{9}{6}$$

$$F_1 = \frac{1}{6}\left(1\cos\frac{\pi}{3} + 2\cos\frac{2\pi}{3} + 3\cos\frac{3\pi}{3} + 2\cos\frac{4\pi}{3} + 1\cos\frac{5\pi}{3}\right)$$

$$= \frac{1}{6}\left(\frac{1}{2} + 2\times\left(-\frac{1}{2}\right) + 3\times(-1) + 2\times\left(-\frac{1}{2}\right) + 1\times\frac{1}{2}\right)$$

$$= \frac{1}{6}\left(\frac{1}{2} - 1 - 3 - 1 + \frac{1}{2}\right) = \frac{1}{6}(-4) = -\frac{4}{6}$$

$$F_2 = \frac{1}{6}\left(1\cos\frac{2\pi}{3} + 2\cos\frac{4\pi}{3} + 3\cos\frac{6\pi}{3} + 2\cos\frac{8\pi}{3} + 1\cos\frac{10\pi}{3}\right)$$

$$= \frac{1}{6}\left(-\frac{1}{2} + 2\times\left(-\frac{1}{2}\right) + 3\times 1 + 2\times\left(-\frac{1}{2}\right) + 1\times\left(-\frac{1}{2}\right)\right)$$

$$= \frac{1}{6}(-1 - 2 + 3) = 0$$

$$F_3 = \frac{1}{6}\left(1\cos\frac{3\pi}{3} + 2\cos\frac{6\pi}{3} + 3\cos\frac{9\pi}{3} + 2\cos\frac{12\pi}{3} + 1\cos\frac{15\pi}{3}\right)$$

$$= \frac{1}{6}(-1 + 2\times 1 + 3\times(-1) + 2\times 1 + 1\times(-1))$$

$$= \frac{1}{6}(-5 + 4) = -\frac{1}{6}$$

$$F_4 = F_2 = 0$$

$$F_5 = F_1 = -\frac{4}{6}.$$

$$\{F_j^2\} = \left\{\frac{9}{4}, \frac{4}{9}, 0, \frac{1}{36}, 0, \frac{4}{9}\right\}.$$

FT($\{h_k\}$):

$$H_0 = \frac{1}{6}\left(\frac{19}{6} + \frac{16}{6} + \frac{11}{6} + \frac{8}{6} + \frac{11}{6} + \frac{16}{6}\right) = \frac{81}{36} = \frac{9}{4}$$

$$H_1 = \frac{1}{6}\left(\frac{19}{6} + \frac{16}{6}\cos\frac{\pi}{3} + \frac{11}{6}\cos\frac{2\pi}{3} + \frac{8}{6}\cos\frac{3\pi}{3} + \frac{11}{6}\cos\frac{4\pi}{3} + \frac{16}{6}\cos\frac{5\pi}{3}\right)$$

$$= \frac{4}{9}$$

$$H_2 = \frac{1}{6}\left(\frac{19}{6} + \frac{16}{6}\cos\frac{2\pi}{3} + \frac{11}{6}\cos\frac{4\pi}{3} + \frac{8}{6}\cos\frac{6\pi}{3} + \frac{11}{6}\cos\frac{8\pi}{3} + \frac{16}{6}\cos\frac{10\pi}{3}\right)$$

$$= 0$$

$$H_3 = \frac{1}{6}\left(\frac{19}{6} + \frac{16}{6}\cos\frac{3\pi}{3} + \frac{11}{6}\cos\frac{6\pi}{3} + \frac{8}{6}\cos\frac{9\pi}{3} + \frac{11}{6}\cos\frac{12\pi}{3} + \frac{16}{6}\cos\frac{15\pi}{3}\right)$$
$$= \frac{1}{36}$$
$$H_4 = H_2 = 0$$
$$H_5 = H_1 = \frac{4}{9}.$$

4.5 Shifting Around

a. The series is even, because of $f_k = +f_{N-k}$.
b. Because of the duality of the forward and inverse transformations (apart from the normalization factor, this only concerns a sign at $e^{-i\omega t} \to e^{+i\omega t}$) the question could also be: Which series produces only a single Fourier coefficient when Fourier-transformed, incidentally at frequency 0? A constant, of course. The Fourier transformation of a "discrete δ-function" therefore is a constant (see Fig. A.10).
c. The series is mixed. It is composed as shown in Fig. A.11.
d. The shifting only results in a phase in F_j, d.h., $|F_j|^2$ stays the same.

Fig. A.10 Answer (b)

Fig. A.11 Answer (c)

Fig. A.12 Real part of the Fourier transform of the random series

$\text{Im}\{F_j\}$

Fig. A.13 Imaginary part of the Fourier transform of the random series

4.6 Pure Noise

a. We get a random series both in the real part (Fig. A.12) and in the imaginary part (Fig. A.13). Random means the absence of any structure.

So all spectral components have to occur, and they, in turn, have to be random, otherwise the inverse transformation would generate a structure.

b. *Trick*: For $N \to \infty$ we can imagine the random series as the discrete version of the function $f(t) = t$ for $-1/2 \le t \le 1/2$. For this purpose we only have to order the numbers of the random series according to their magnitudes! According to Parseval's theorem (4.31) we don't have to do a Fourier transformation at all. So with $2N + 1$ samples we need:

$$\frac{2}{2N+1} \sum_{k=0}^{N} \left(\frac{k}{N}\right)^2 = \frac{2}{2N+1} \frac{1}{4N^2} \frac{(2N+1)N(N+1)}{6} \tag{A.1}$$

$$= \frac{N+1}{12N}; \qquad \lim_{N \to \infty} \frac{N+1}{12N} = \frac{1}{12}.$$

We could have solved the following integral instead:

$$\int_{-0.5}^{+0.5} t^2 \mathrm{d}t = 2 \int_{0}^{+0.5} t^2 \mathrm{d}t = 2\frac{t^3}{3} \Big|_0^{0.5} = \frac{2}{3}\frac{1}{8} = \frac{1}{12}. \tag{A.2}$$

Let's compare: $0.5 \cos \omega t$ has, due to $\overline{\cos^2 \omega t} = 0.5$, the signal power $0.5^2 \times 0.5 = 1/8$.

$\text{Re}\{F_j\}$

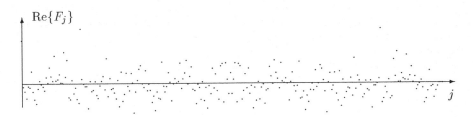

Fig. A.14 Real part of the Fourier transform according to (4.58)

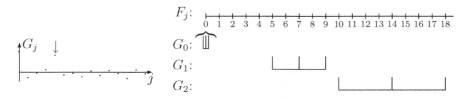

Fig. A.15 Result of the cross correlation: at the position of the fundamental frequency at channel 4 the "signal" (arrow) is clearly visible; channel 0 also happens to run up, however, there is no corresponding pattern

4.7 Pattern Recognition

It's best to use the cross correlation. It is formed with the Fourier transform of the experimental data Fig. A.14 and the theoretical "frequency comb", the pattern (Fig. 4.29). As we're looking for cosine patterns, we only use the real part for the cross correlation.

Here, channel 36 goes up (from 128 channels to Ω_{Nyq}). The right half is the mirror image of the left half. So the Fourier transform suggests only a spectral component (apart from noise) at $(36/128)\Omega_{\text{Nyq}} = (9/32)\Omega_{\text{Nyq}}$. If we search for pattern Fig. 4.29 in the data, we get something totally different.

The result of the cross correlation with the theoretical frequency comb leads to the following algorithm:

$$G_j = F_{5j} + F_{7j} + F_{9j}. \tag{A.3}$$

The result shows Fig. A.15.

So the noisy signal contains cosine components with the frequencies $5\pi(4/128)$, $7\pi(4/128)$, and $9\pi(4/128)$.

4.8 Go on the Ramp (for Gourmets only)

The series is mixed because neither $f_k = f_{-k}$ nor $f_k = -f_{-k}$ is true.

Decomposition into even and odd parts.

We have the following equations:

$$\begin{aligned} f_k &= f_k^{\text{even}} + f_k^{\text{odd}} \\ f_k^{\text{even}} &= f_{N-k}^{\text{even}} \qquad \text{for } k = 0, 1, \ldots, N-1. \\ f_k^{\text{odd}} &= -f_{N-k}^{\text{odd}} \end{aligned}$$

The first condition gives N equations for $2N$ unknowns. The second and third equations give each N further conditions, each appears twice, hence we have N additional equations. Instead of solving this system of linear equations, we solve the problem by arguing.

First, because of $f_0^{\text{odd}} = 0$ we have $f_0^{\text{even}} = 0$. Shifting the ramp downwards by $N/2$ we already have an odd function with the exception of $k = 0$ (see Fig. A.16):

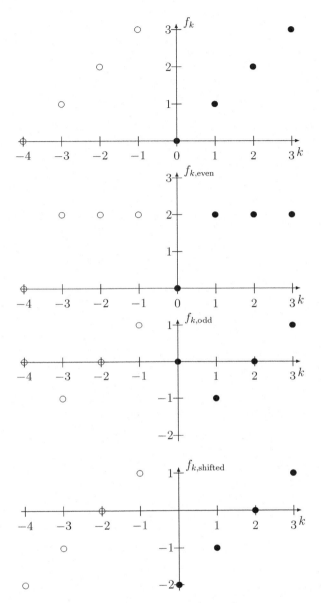

Fig. A.16 One-sided ramp for $N = 4$ (periodic continuation with open circles); decomposition into even and odd parts; ramp shifted downwards by 2 immediately gives the odd part (except for $k = 0$) (*from top to bottom*)

$$f_k^{\text{shifted}} = k - \frac{N}{2} \quad \text{for } k = 0, 1, 2, \ldots, N - 1.$$

$$f_{-k}^{\text{shifted}} = f_{N-k}^{\text{shifted}} = (N - k) - \frac{N}{2} = \frac{N}{2} - k$$

$$= -\left(k - \frac{N}{2}\right).$$

So we have already found the odd part:

$$f_k^{\text{odd}} = k - \frac{N}{2} \quad \text{for } k = 1, 2, \ldots, N - 1$$

$$f_0^{\text{odd}} = 0$$

and, of course, we have also found the real part:

$$f_k^{\text{even}} = \frac{N}{2} \quad \text{for } k = 1, 2, \ldots, N - 1 \quad \text{(compensates for the shift)}$$

$$f_0^{\text{even}} = 0 \quad \text{(see above)}.$$

Real part of Fourier transform:

$$\text{Re}\{F_j\} = \frac{1}{N} \sum_{k=1}^{N-1} \frac{N}{2} \cos \frac{2\pi k j}{N}.$$

Dirichlet: $1/2 + \cos x + \cos 2x + \cdots + \cos Nx = \sin[(N + 1/2)x]/(2\sin[x/2])$; here we have $x = 2\pi j/N$ and instead of N we go until $N - 1$:

$$\sum_{k=1}^{N-1} \cos kx = \frac{\sin(N - \frac{1}{2})x}{2\sin \frac{x}{2}} - \frac{1}{2}$$

$$= \frac{\overset{=0}{\sin Nx \cos \frac{x}{2}} - \overset{=1}{\cos Nx} \sin \frac{x}{2}}{2\sin \frac{x}{2}} - \frac{1}{2}$$

$$= -\frac{1}{2} - \frac{1}{2} = -1.$$

$$\text{Re}\{F_0\} = \frac{1}{N} \frac{N}{2} \underbrace{(N - 1)}_{\text{number of terms}} = \frac{N-1}{2}, \qquad \text{Re}\{F_j\} = -\frac{1}{2}.$$

Check:

$$\text{Re}\{F_0\} + \sum_{j=1}^{N-1} \text{Re}\{F_j\} = \frac{N-1}{2} - \frac{1}{2}(N - 1) = 0.$$

Imaginary part of Fourier transform:

$$\text{Im}\{F_j\} = \frac{1}{N} \sum_{k=1}^{N-1} \left(k - \frac{N}{2} \right) \sin \frac{2\pi k j}{N}.$$

For the sum over sines we need the analogue of Dirichlet's kernel for sines. Let us try an expression with an unknown numerator but the same denominator as for the sum of cosines:

$$\sin x + \sin 2x + \cdots + \sin(N-1)x = \frac{?}{2 \sin \frac{x}{2}}$$

$$2 \sin \frac{x}{2} \sin x + 2 \sin \frac{x}{2} \sin 2x + \ldots + 2 \sin \frac{x}{2} \sin(N-1)x$$

$$= \underbrace{\cos \frac{x}{2} - \cos \frac{3x}{2} + \cos \frac{3x}{2}}_{=0}$$

$$\underbrace{- \cos \frac{5x}{2} + \cdots + \cos \left(N - \frac{3}{2} \right) x}_{=0}$$

$$- \cos \left(N - \frac{1}{2} \right) x$$

$$= \cos \frac{x}{2} - \cos \left(N - \frac{1}{2} \right) x$$

$$\longrightarrow \quad \sum_{k=1}^{N-1} \sin kx = \frac{\cos \frac{x}{2} - \cos \left(N - \frac{1}{2} \right) x}{2 \sin \frac{x}{2}}$$

$$= \frac{\cos \frac{x}{2} - \overset{=1}{\cos Nx} \cos \frac{x}{2} + \overset{=0}{\sin Nx} \sin \frac{x}{2}}{2 \sin \frac{x}{2}} = 0.$$

Hence, there remains only the term with $k \sin(2\pi k j/N)$. We can evaluate this sum by differentiating the formula for Dirichlet's kernel (Use the general formula and insert $x = 2\pi j/N$ into the differentiated formula!):

$$\frac{\mathrm{d}}{\mathrm{d}x} \sum_{k=1}^{N-1} \cos kx = - \sum_{k=1}^{N-1} k \sin kx$$

$$= \frac{1}{2} \frac{\left(N - \frac{1}{2} \right) \cos \left[\left(N - \frac{1}{2} \right) x \right] \sin \frac{x}{2} - \sin \left[\left(N - \frac{1}{2} \right) x \right] \frac{1}{2} \cos \frac{x}{2}}{\sin^2 \frac{x}{2}}$$

$$= \frac{1}{2} \frac{\left(N - \frac{1}{2} \right) \left(\overset{=1}{\cos Nx} \cos \frac{x}{2} \right) \sin \frac{x}{2} - \left(\overset{=0}{\sin Nx} \cos \frac{x}{2} - \overset{=1}{\cos Nx} \sin \frac{x}{2} \right) \frac{1}{2} \cos \frac{x}{2}}{\sin^2 \frac{x}{2}}$$

$$= \frac{1}{2}\left(\left(N - \frac{1}{2}\right)\frac{\cos\frac{x}{2}}{\sin\frac{x}{2}} + \frac{1}{2}\frac{\cos\frac{x}{2}}{\sin\frac{x}{2}}\right)$$

$$= \frac{N}{2}\frac{\cos\frac{x}{2}}{\sin\frac{x}{2}} = \underline{\underline{\frac{N}{2}\cot\frac{\pi j}{N}}}$$

$$\mathrm{Im}\{F_j\} = \frac{1}{N}(-1)\frac{N}{2}\cot\frac{\pi j}{N} = -\frac{1}{2}\cot\frac{\pi j}{N}, \qquad j \neq 0, \qquad \mathrm{Im}\{F_0\} = 0,$$

finally together:

$$F_j = \begin{cases} -\dfrac{1}{2} - \dfrac{i}{2}\cot\dfrac{\pi j}{N} & \text{for } j \neq 0 \\[3mm] \dfrac{N-1}{2} & \text{for } j = 0 \end{cases}.$$

Parseval's theorem:

left hand side: $\dfrac{1}{N}\displaystyle\sum_{k=1}^{N-1} k^2 = \dfrac{1}{N}\dfrac{(N-1)N(2(N-1)+1)}{6} = \dfrac{(N-1)(2N-1)}{6}$

right hand side: $\left(\dfrac{N-1}{2}\right)^2 + \dfrac{1}{4}\displaystyle\sum_{j=1}^{N-1}\left(1 + i\cot\dfrac{\pi j}{N}\right)\left(1 - i\cot\dfrac{\pi j}{N}\right)$

$$= \left(\frac{N-1}{2}\right)^2 + \frac{1}{4}\sum_{j=1}^{N-1}\left(1 + \cot^2\frac{\pi j}{N}\right)$$

$$= \left(\frac{N-1}{2}\right)^2 + \frac{1}{4}\sum_{j=1}^{N-1}\frac{1}{\sin^2\frac{\pi j}{N}}$$

hence:

$$\frac{(N-1)(2N-1)}{6} = \left(\frac{N-1}{2}\right)^2 + \frac{1}{4}\sum_{j=1}^{N-1}\frac{1}{\sin^2\frac{\pi j}{N}}$$

or $\dfrac{1}{4}\displaystyle\sum_{j=1}^{N-1}\dfrac{1}{\sin^2\frac{\pi j}{N}} = \dfrac{(N-1)(2N-1)}{6} - \dfrac{(N-1)^2}{4}$

$$= (N-1)\frac{(2N-1)2 - (N-1)3}{12}$$

$$= \frac{N-1}{12}(4N - 2 - 3N + 3)$$

$$= \frac{N-1}{12}(N+1) = \frac{N^2-1}{12}$$

and finally:

$$\sum_{j=1}^{N-1} \frac{1}{\sin^2 \frac{\pi j}{N}} = \frac{N^2 - 1}{3}.$$

The result for $\sum_{j=1}^{N-1} \cot^2(\pi j/N)$ is obtained as follows: we use Parseval's theorem for the real/even and imaginary/odd parts separately. For the real part we get:

left hand side: $\dfrac{1}{N} \left(\dfrac{N}{2}\right)^2 (N-1) = \dfrac{N(N-1)}{4}$

right hand side: $\left(\dfrac{N-1}{2}\right)^2 + \dfrac{N-1}{4} = \dfrac{N(N-1)}{4}.$

The real parts are equal, so the imaginary parts of the left and right hand sides have to be equal, too.

For the imaginary part we get:

left hand side:

$$\frac{1}{N} \sum_{k=1}^{N-1} \left(\frac{k-N}{2}\right)^2 = \frac{1}{N} \sum_{k=1}^{N-1} \left(k^2 - kN + \frac{N^2}{4}\right)$$

$$= \frac{1}{N} \left(\frac{(N-1)N(2N-1)}{6} - \frac{N(N-1)N}{2} + \frac{N^2(N-1)}{4}\right)$$

$$= \frac{(N-1)(N-2)}{12}$$

right hand side:

$$\frac{1}{4} \sum_{j=1}^{N-1} \cot^2 \frac{\pi j}{N}$$

from which we get $\sum_{j=1}^{N-1} \cot^2 \frac{\pi j}{N} = (N-1)(N-2)/3.$

4.9 Transcendental (for Gourmets only)

The series is even because:

$$f_{-k} = f_{N-k} \stackrel{?}{=} f_k.$$

Insert $N - k$ into (4.59) on both sides:

$$f_{N-k} = \begin{cases} N-k & \text{for } N-k = 0, 1, \ldots, N/2-1 \\ N-(N-k) & \text{for } N-k = N/2, N/2+1, \ldots, N-1 \end{cases}$$

$$\text{or} \quad f_{N-k} = \begin{cases} N-k & \text{for } k = N, N-1, \ldots, N/2+1 \\ k & \text{for } k = N/2, N/2-1, \ldots, 1 \end{cases}$$

$$\text{or} \quad f_{N-k} = \begin{cases} k & \text{for } k = 1, 2, \ldots, N/2 \\ N-k & \text{for } k = N/2+1, \ldots, N \end{cases},$$

a. for $k = N$ we have $f_0 = 0$, so we could include it also in the first line because $f_N = f_0 = 0$.

b. for $k = N/2$ we have $f_{N/2} = N/2$, so we could include it also in the second line.

This completes the proof. Since the series is even, we only have to calculate the real part:

$$F_j = \frac{1}{N} \sum_{k=0}^{N-1} f_k \cos \frac{2\pi k j}{N}$$

$$= \frac{1}{N} \left(\sum_{k=0}^{\frac{N}{2}-1} k \cos \frac{2\pi k j}{N} + \sum_{k=\frac{N}{2}}^{N-1} (N-k) \cos \frac{2\pi k j}{N} \right) \qquad \text{with } k' = N - k$$

$$= \frac{1}{N} \left(\sum_{k=0}^{\frac{N}{2}-1} k \cos \frac{2\pi k j}{N} + \sum_{k'=\frac{N}{2}}^{1} k' \cos \frac{2\pi (N-k') j}{N} \right)$$

$$= \frac{1}{N} \left(\sum_{k=0}^{\frac{N}{2}-1} k \cos \frac{2\pi k j}{N} \right.$$

$$\left. + \sum_{k'=1}^{\frac{N}{2}} k' \left(\underbrace{\cos \frac{2\pi N j}{N}}_{=1} \cos \frac{2\pi k' j}{N} + \underbrace{\sin \frac{2\pi N j}{N}}_{=0} \sin \frac{2\pi (-k') j}{N} \right) \right)$$

$$= \frac{1}{N} \left(\sum_{k=0}^{\frac{N}{2}-1} k \cos \frac{2\pi k j}{N} + \sum_{k'=1}^{\frac{N}{2}} k' \cos \frac{2\pi k' j}{N} \right)$$

$$= \frac{1}{N} \left(2 \sum_{k=1}^{\frac{N}{2}-1} k \cos \frac{2\pi k j}{N} + \frac{N}{2} \cos \pi j \right) \qquad \text{with } \frac{2\pi \frac{N}{2} j}{N} = \pi j$$

$$= \frac{2}{N} \sum_{k=1}^{\frac{N}{2}-1} k \cos \frac{2\pi k j}{N} + \frac{1}{2} (-1)^j.$$

This can be simplified further.

How can we get this sum? Let us try an expression with an unknown numerator but the same denominator as for the sum of cosines ("sister" analogue of Dirichlet's kernel):

$$\sum_{k=1}^{\frac{N}{2}-1} \sin kx = \frac{?}{2 \sin \frac{x}{2}} \qquad \text{with } x = \frac{2\pi j}{N}.$$

The numerator of the right hand side is:

$$2 \sin \frac{x}{2} \sin x + 2 \sin \frac{x}{2} \sin 2x + \cdots + 2 \sin \frac{x}{2} \sin \left(\frac{N}{2} - 1\right) x$$

$$= \underbrace{\cos \left(\frac{x}{2}\right) - \cos \left(\frac{3x}{2}\right) + \cos \left(\frac{3x}{2}\right) - \cdots}_{=0}$$

$$\underbrace{- \cos \left(\frac{N}{2} - \frac{3}{2}\right) x + \cos \left(\frac{N}{2} - \frac{3}{2}\right) x - \cos \left(\frac{N}{2} - \frac{1}{2}\right) x}_{=0}$$

$$= \cos \frac{x}{2} - \cos \frac{N-1}{2} x.$$

Finally we get:

$$\sum_{k=1}^{\frac{N}{2}-1} \sin kx = \frac{\cos \frac{x}{2} - \cos \frac{N-1}{2} x}{2 \sin \frac{x}{2}}, \quad N = \text{even, do not use for } x = 0.$$

Now we take the derivative with respect to x. Let us exclude the special case of $x = 0$. We shall treat it later.

$$\frac{d}{dx} \sum_{k=1}^{\frac{N}{2}-1} \sin kx = \sum_{k=1}^{\frac{N}{2}-1} k \cos kx$$

$$= \frac{1}{2} \frac{\left[-\frac{1}{2} \sin \frac{x}{2} + \left(\frac{N-1}{2}\right) \sin \left(\frac{N-1}{2}\right) x\right] \sin \frac{x}{2} - \left[\cos \frac{x}{2} - \cos \left(\frac{N-1}{2}\right) x\right] \frac{1}{2} \cos \frac{x}{2}}{\sin^2 \frac{x}{2}}$$

$$= \frac{1}{2} \frac{-\frac{1}{2} \sin^2 \frac{x}{2} - \frac{1}{2} \cos^2 \frac{x}{2} + \left(\frac{N-1}{2}\right) \overbrace{\left(\sin \frac{Nx}{2} \cos \frac{x}{2} - \cos \frac{Nx}{2} \sin \frac{x}{2}\right)}^{=0} \sin \frac{x}{2}}{\sin^2 \frac{x}{2}}$$

$$+\frac{1}{2}\left(\cos\frac{Nx}{2}\cos\frac{x}{2}+\overset{=0}{\sin\frac{Nx}{2}}\sin\frac{x}{2}\right)\cos\frac{x}{2}$$

with $x=\dfrac{2\pi j}{N}$, $\cos\dfrac{Nx}{2}=\cos\pi j=(-1)^{j}$, $\sin\dfrac{Nx}{2}=\sin\pi j=0$

$$=\frac{1}{2}\frac{-\dfrac{1}{2}+\dfrac{N-1}{2}(-1)^{j+1}\sin^{2}\dfrac{x}{2}+\dfrac{1}{2}(-1)^{j}\cos^{2}\dfrac{x}{2}}{\sin^{2}\dfrac{x}{2}}$$

$$=\frac{1}{2}\frac{-\dfrac{1}{2}+(-1)^{j+1}\dfrac{N}{2}\sin^{2}\dfrac{x}{2}-\dfrac{1}{2}(-1)^{j}\overset{=-1}{\left(-\cos^{2}\dfrac{x}{2}-\sin^{2}\dfrac{x}{2}\right)}}{\sin^{2}\dfrac{x}{2}}$$

$$=\frac{1}{2}\left(\frac{1}{2\sin^{2}\dfrac{x}{2}}\left((-1)^{j}-1\right)+(-1)^{j+1}\dfrac{N}{2}\right)$$

$$\Rightarrow F_{j}=\frac{2}{N}\left(\frac{(-1)^{j}-1}{2}\frac{1}{2}\frac{1}{\sin^{2}\dfrac{\pi j}{N}}+(-1)^{j+1}\dfrac{N}{4}\right)+\frac{1}{2}(-1)^{j}$$

$$=\frac{(-1)^{j}-1}{2N\sin^{2}\dfrac{\pi j}{N}}$$

$$=\begin{cases}-\dfrac{1}{N\sin^{2}\dfrac{\pi j}{N}} & \text{for } j=\text{odd}\\ 0 & \text{else}\end{cases}.$$

The special case of $j=0$ is obtained from:

$$\sum_{k=1}^{\frac{N}{2}-1}k=\frac{\left(\dfrac{N}{2}-1\right)\dfrac{N}{2}}{2}=\frac{N^{2}}{8}-\frac{N}{4}.$$

Hence:

$$F_{0}=\frac{2}{N}\left(\frac{N^{2}}{8}-\frac{N}{4}\right)+\frac{1}{2}=\frac{N}{4}.$$

We finally have:

$$F_{j}=\begin{cases}-\dfrac{1}{N\sin^{2}\dfrac{\pi j}{N}} & \text{for } j=\text{odd}\\ 0 & \text{for } j=\text{even}, \ j\neq 0\\ \dfrac{N}{4} & \text{for } j=0\end{cases}.$$

Now we use Parseval's theorem:

l.h.s.
$$\frac{1}{N}\left[2\frac{\left(\frac{N}{2}-1\right)\frac{N}{2}\left(2\left(\frac{N}{2}-1\right)+1\right)}{6}+\frac{N^2}{4}\right]$$

$$=\frac{1}{N}\left[2\frac{1}{2}\frac{(N-2)\frac{1}{2}N(N-1)}{6}+\frac{N^2}{4}\right]$$

$$=\frac{1}{N}\left[\frac{N(N-1)(N-2)+3N^2}{12}\right]=\frac{(N-1)(N-2)+3N}{12}$$

$$=\underline{\underline{\frac{N^2+2}{12}}}$$

r.h.s.
$$\frac{N^2}{16}+\sum_{\substack{j=1\\\text{odd}}}^{N-1}\frac{1}{N^2\sin^4\frac{\pi j}{N}}\qquad\text{with } j=2k-1$$

$$=\sum_{k=1}^{N/2}\frac{1}{N^2\sin^4\frac{\pi(2k-1)}{N}}+\frac{N^2}{16}$$

which gives:

$$\frac{N^2}{12}+\frac{1}{6}=\sum_{k=1}^{N/2}\frac{1}{N^2\sin^4\frac{\pi(2k-1)}{N}}+\frac{N^2}{16}$$

and finally:

$$\sum_{k=1}^{N/2}\frac{1}{\sin^4\frac{\pi(2k-1)}{N}}=\frac{N^2(N^2+8)}{48}.$$

The right hand side can be shown to be an integer! Let $N=2M$.

$$\frac{4M^2(4M^2+8)}{48}=\frac{4M^24(M^2+2)}{48}=\frac{M^2(M^2+2)}{3}$$

$$=\frac{M(M-1)M(M+1)+3M^2}{3}$$

$$=M\frac{(M-1)M(M+1)}{3}+M^2.$$

Three consecutive numbers can always be divided by 3!

Now we use the high-pass property:

$$\sum_{j=0}^{N-1} F_j = \frac{N}{4} - \frac{1}{N} \sum_{\substack{j=1 \\ \text{odd}}}^{N-1} \frac{1}{\sin^2 \dfrac{\pi j}{N}} \quad \text{with } j = 2k - 1$$

$$= \frac{N}{4} - \frac{1}{N} \sum_{k=1}^{\frac{N}{2}} \frac{1}{\sin^2 \dfrac{\pi(2k-1)}{N}}.$$

For a high-pass filter we must have $\sum_{j=0}^{N-1} F_j = 0$ because a zero frequency must not be transmitted (see Chap. 5). If you want, use definition (4.13) with $k = 0$ and interpret f_k being the filter in the frequency domain and F_j its Fourier transform. Hence, we get:

$$\sum_{k=1}^{N/2} \frac{1}{\sin^2 \dfrac{\pi(2k-1)}{N}} = \frac{N^2}{4}.$$

Since N is even, the result is always integer!

These are nice examples how a finite sum over an expression involving a transcendental function yields an integer!

Playground of Chap. 5

5.1 Totally Different

The first central difference is:

"exact"

$$y_k = \frac{f_{k+1} - f_{k-1}}{2\Delta t} \qquad\qquad f'(t) = -\frac{\pi}{2} \sin \frac{\pi}{2} t$$

$$y_0 = \frac{f_1 - f_{-1}}{2/3} = \frac{f_1 - f_5}{2/3} = \frac{1 + \sqrt{3}/2}{2/3} = 2.799 \quad f'(t_0) = 0$$

$$y_1 = \frac{f_2 - f_0}{2/3} = \frac{1/2 - 1}{2/3} = -0.750 \qquad f'(t_1) = -\frac{\pi}{2} \sin \frac{\pi}{2}\frac{1}{3} = -0.7854$$

$$y_2 = \frac{f_3 - f_1}{2/3} = \frac{0 - \sqrt{3}/2}{2/3} = -1.299 \qquad f'(t_2) = -\frac{\pi}{2} \sin \frac{\pi}{2}\frac{2}{3} = -1.3603$$

$$y_3 = \frac{f_4 - f_2}{2/3} = \frac{-1/2 - 1/2}{2/3} = -1.500 \qquad f'(t_3) = -\frac{\pi}{2} \sin \frac{\pi}{2}\frac{3}{3} = -1.5708$$

$$y_4 = \frac{f_5 - f_3}{2/3} = \frac{-\sqrt{3}/2 - 0}{2/3} = -1.299 \qquad f'(t_4) = -\frac{\pi}{2} \sin \frac{\pi}{2}\frac{4}{3} = -1.3603$$

$$y_5 = \frac{f_6 - f_4}{2/3} = \frac{f_0 - f_4}{2/3} = \frac{1 + 1/2}{2/3} = 2.250 \quad f'(t_5) = -\frac{\pi}{2} \sin \frac{\pi}{2}\frac{5}{3} = -0.7854.$$

Of course, the beginning y_0 and the end y_5 are totally wrong because of the periodic continuation. Let us calculate the relative error for the other derivatives:

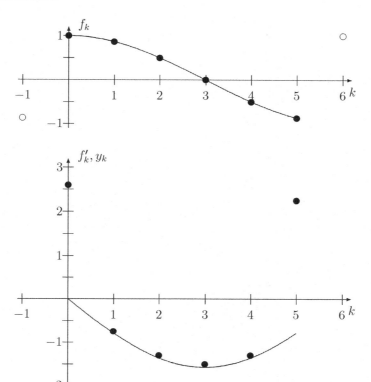

Fig. A.17 Input $f_k = \cos \pi t_k/2$, $t_k = k \Delta t$ with $k = 0, 1, \dots, 5$ and $\Delta t = 1/3$ (*top*). First central difference (*bottom*). The solid line is the exact derivative. y_0 and y_5 appear to be totally wrong. However, we must not forget the periodic continuation of the series (*see open circles in the top panel*)

$$k = 1 \quad \frac{\text{exact} - \text{discrete}}{\text{exact}} = \frac{-0.7854 + 0.750}{-0.7854} = 4.5\,\% \text{ too small}$$

$k = 2 \quad 4.5\,\%$ too small

$k = 3 \quad 4.5\,\%$ too small

$k = 4 \quad 4.5\,\%$ too small.

The result is plotted in Fig. A.17.

5.2 Simpson's-1/3 Versus Trapezoid

The exact, trapezoidal, and Simpson's-1/3 calculations are illustrated in Fig. A.18.

Trapezoid:

$$I = \left(\frac{f_0}{2} + \sum_{k=1}^{3} f_k + \frac{f_4}{2} \right)$$

$$= \left(\frac{1}{2} + 0.5 - 0.5 - 1 - \frac{0.5}{2} \right) = -0.75,$$

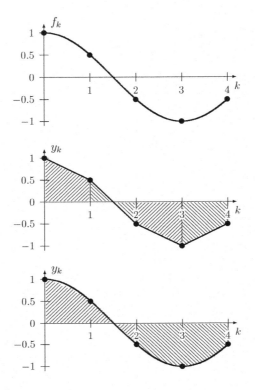

Fig. A.18 Input $f_k = \cos \pi t_k$, $t_k = k\Delta t$, $k = 0, 1, \ldots, 4$, $\Delta t = 1/3$ (*top*). Area of trapezoids to be added up. Step width is Δt (*middle*). Area of parabolically interpolated segment in Simpson's 1/3-rule. Step width is $2\Delta t$ (*bottom*)

Simpson's-1/3:

$$
I = \left(\frac{f_2 + 4f_1 + f_0}{3} \right) + \left(\frac{f_4 + 4f_3 + f_2}{3} \right)
$$
$$
= \left(\frac{-0.5 + 4 \times 0.5 + 1}{3} \right) + \left(\frac{-0.5 + 4 \times (-1) + (-0.5)}{3} \right) = -0.833.
$$

In order to derive the exact value we have to convert $f_k = \cos(k\pi\Delta t/3)$ into $f(t) = \cos(\pi t/3)$. Hence, we have $\int_0^4 \cos(\pi t/3)\mathrm{d}t = -0.82699$.

The relative errors are:

$$
1 - \frac{\text{trapezoid}}{\text{exact}} = 1 - \frac{-0.75}{-0.82699} \Rightarrow 9.3\,\% \text{ too small,}
$$
$$
1 - \frac{\text{Simpson's-1/3}}{\text{exact}} = 1 - \frac{-0.833}{-0.82699} \Rightarrow 0.7\,\% \text{ too large.}
$$

This is consistent with the fact that the Trapezoidal Rule always underestimates the integral whereas Simpson's 1/3-rule always overestimates (see Figs. 5.14 and 5.15).

5.3 Totally Noisy

a. You get random noise, and additionally in the real part (because of the cosine!), a discrete line at frequency $(1/4)\Omega_{Nyq}$ (see Figs. A.19 and A.20).
b. If you process the input using a simple low-pass filter (5.12), the time signal already looks better as shown in Fig. A.21. The real part of the Fourier transform of the filtered function is shown in Fig. A.22.

Fig. A.19 Real part of the Fourier transform of the series according to (5.47)

Fig. A.20 Imaginary part of the Fourier transform of the series according to (5.47)

Fig. A.21 Input that has been processed using a low-pass filter according to (5.47)

Fig. A.22 Real part of the Fourier transform of the filtered function y_k according to Fig. A.21

5.4 Inclined Slope

a. We simply use a high-pass filter (cf. (5.13)). The result is shown in Fig. A.23.
b. For a "δ-shaped line" as input we get precisely the definition of the high-pass filter as result. This leads to the following recommendation for a high-pass filter with smaller undershoots:

$$y_k = \frac{1}{8}(-f_{k-2} - f_{k-1} + 4f_k - f_{k+1} - f_{k+2}). \tag{A.4}$$

The result of this data processing is shown in Fig. A.24. If we keep going, we'll easily recognise Dirichlet's integral kernel (1.53), that belongs to a step. The problem here is that boundary effects are progressively harder to handle. Using recursive filters, naturally, is much better suited to processing data.

Fig. A.23 Data from Fig. 5.17 processed using the high-pass filter $y_k = (1/4)(-f_{k-1}+2f_k-f_{k+1})$. The "undershoots" don't look very good

Fig. A.24 Data according to Fig. 5.17, processed with the modified high-pass filter according to (A.4). The undershoots get a bit smaller and wider. Progress admittedly is small, yet visible

Playground of Chap. 6

6.1 What's Your Average?
The transfer function of the Lagrange interpolator with $N = 1$ is unity at zero frequency, hence the average does not depend on d. In case of doubt, calculate it by "brute force".

6.2 Late Impulse
See Figs. A.25 and A.26.

6.3 The Devil Takes the Hindmost

a. See Fig. A.27.
b. The undershoot at the leading edge is compensated at the trailing edge, as are all other deviations from unity.
c. None of the spectral components is attenuated, i.e. nothing of the input gets lost, the frequency-dependent group delay is responsible for the fact that it merely gets redistributed. I am sure you have noticed that the Thiran all-pass filter can be used to sum up certain infinite series.

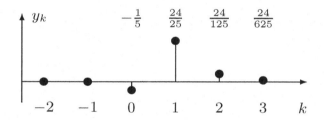

Fig. A.25 Response of the $N = 1$ Thiran all-pass filter to an impulse with $d = 1/2$

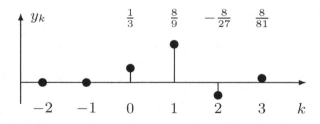

Fig. A.26 Response of the $N = 1$ Thiran all-pass filter to an impulse with $d = -1/2$

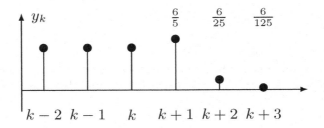

Fig. A.27 Response of the $N = 1$ Thiran all-pass filter to the trailing edge of a unit pulse for $d = 1/2$

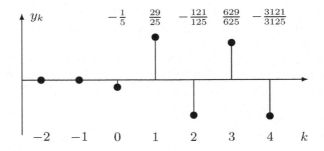

Fig. A.28 Response of the $N = 1$ Thiran all-pass filter to an input with $\cos \Omega_{\text{Nyq}}$ starting at $t = 0$ and being 0 at earlier times for $d = 1/2$

6.4 Delayed Nyquist

Apart from a transient, the full amplitude of $\cos \Omega_{\text{Nyq}}$ is approached rapidly. The Lagrange $N = 1$ interpolator would not transmit the signal at all. See Fig. A.28.

Playground of Chap. 7

7.1 Go on the ramp, not on the rampage!

a. First Shifting Rule

The coefficients of the Fourier transform of the triangular function for a shift by M are multiplied by $W_{2M}^{-jM} = \cos \pi j = -1$ for $j = 1, 3, \ldots$ with the exception of $F_0 = 1/2$ which remains unchanged. Multiplying the result by $2M$ we get:

$$
F_j = \begin{cases} -\dfrac{1}{M} \sin^2 \dfrac{\pi j}{2M} & \text{for odd } j \\[2ex] 0 & \text{for even } j \neq 0 \end{cases} .
$$

b. Tricky Carl Friedrich Gauß

The double-sided ramp is 1 minus the triangular function. Hence, all non-zero Fourier coefficients except $j = 0$ are the same as for the triangular function but negative and we have to multiply by $2M$ to get $F_0 = 1 - 1/2 = 1/2$. The result is identical, of course:

$$
F_j = \begin{cases} -\dfrac{1}{M} \sin^2 \dfrac{\pi j}{2M} & \text{for odd } j \\[2ex] 0 & \text{for even } j \end{cases}.
$$

After all, the triangular function and the double-sided ramp complement each other like man and woman.

7.2 Slice it!

$P(x) = \int_{-\infty}^{+\infty} \rho(x, y)\mathrm{d}y$.

$\mathrm{FT}(k_x, k_y) = \iint_{-\infty}^{+\infty} \rho(x, y)\mathrm{e}^{-i(k_x x + k_y y)}\mathrm{d}x\mathrm{d}y$.

For the "central slice" we get:

$$
\begin{aligned}
\mathrm{FT}(k_x, 0) &= \iint_{-\infty}^{+\infty} \rho(x, y)\mathrm{e}^{-ik_x x}\mathrm{d}x\mathrm{d}y \\
&= \int_{-\infty}^{+\infty} \left(\int_{-\infty}^{+\infty} \rho(x, y)\mathrm{d}y \right) \mathrm{e}^{-ik_x x}\mathrm{d}x \\
&= \int_{-\infty}^{+\infty} P(x)\mathrm{e}^{-ik_x x}\mathrm{d}x.
\end{aligned}
$$

This is the 1D-Fourier transform of $P(x)$.

7.3 Reconstruct it!

The inverse FT of double-sided ramp filter: ($N = 2$)

$$
g_0 = (G_0 + G_1) = 1
$$
$$
g_1 = (G_0 + G_1 \mathrm{e}^{i\pi}) = -1
$$

The convolution is defined as follows:

$$
h_k = \frac{1}{2} \sum_{l=0}^{1} f_l g_{k-l}.
$$

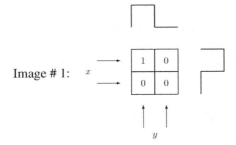

Image # 1:

Convolution:
Note, that the f's are the projections.

x-direction: $f_0 = 1 \quad f_1 = 0$

$$h_0 = \frac{1}{2}(f_0 g_0 + f_1 g_1) = \frac{1}{2}(1 \times 1 + 0 \times (-1)) = +\frac{1}{2}$$

$$h_1 = \frac{1}{2}(f_0 g_1 + f_1 g_0) = \frac{1}{2}(1 \times (-1) + 0 \times 1) = -\frac{1}{2}$$

y-direction: $f_0 = 1 \quad f_1 = 0$, hence, we get the same result as for the x-direction

$$h_0 = +\frac{1}{2}$$

$$h_1 = -\frac{1}{2}$$

convolved: backprojected:

The box with -1 is an reconstruction artefact. Use a cutoff: all negative values
do not correspond to an object.

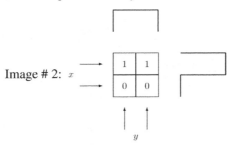

Image # 2:

Convolution:

$$x\text{-direction:} \quad f_0 = 2 \quad f_1 = 0$$
$$h_0 = \frac{1}{2}(2 \times 1 + 0 \times (-1)) = +1$$
$$h_1 = \frac{1}{2}(2 \times (-1) + 0 \times 1) = -1$$
$$y\text{-direction:} \quad f_0 = 1 \quad f_1 = 1$$
$$h_0 = \frac{1}{2}(1 \times 1 + 1 \times (-1)) = 0$$
$$h_1 = \frac{1}{2}(1 \times (-1) + 1 \times 1) = 0$$

convolved: backprojected:

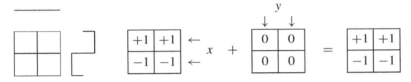

Here, we have an interesting situation: the filtered y-projection vanishes identically because a constant—don't forget the periodic continuation—cannot pass through a high-pass filter. In other words, a uniform object looks like no object at all! All that matters is contrast!

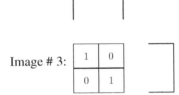

Image # 3:

This "diagonal object" cannot be reconstructed. We would require projections along the diagonals!

Image # 4: $\begin{array}{|c|c|}\hline 1 & 1 \\\hline 1 & 0 \\\hline\end{array}$ is the "reverse" of image # 1.

Image # 5: $\begin{array}{|c|c|}\hline 1 & 1 \\\hline 1 & 1 \\\hline\end{array}$ is like a white rabbit in snow or a black panther in the dark.

References

1. H.J. Lipkin, *Beta-Decay for Pedestrians* (North-Holland Publ, Amsterdam, 1962)
2. H.J. Weaver, *Applications of Discrete and Continuous Fourier Analysis* (A Wiley-Interscience Publication, New York, 1983)
3. H.J. Weaver, *Theory of Discrete and Continuous Fourier Analysis* (Wiley, New York, 1989)
4. T. Butz, *Fouriertransformation für Fußgänger* (Teubner, Wiesbaden, 2003)
5. Wolfram Research, Inc., Mathematica, Version 10.1, Champaign, IL, USA
6. E. Zeidler (ed.), *Oxford Users' Guide to Mathematics* (Oxford University Press, Oxford, 2004)
7. W.H. Press, B.P. Flannery, S.A. Teukolsky, W.T. Vetterling, *Numerical Recipes, The Art of Scientific Computing* (Cambridge University Press, New York, 1989)
8. I.S. Gradshteyn, I.M. Ryzhik, *Tables of Integrals, Series, and Products* (Academic Press Inc, San Diego, 1980)
9. F.J. Harris, Proc. IEEE **66**, 51 (1978)
10. M. Abramowitz, I.A. Stegun, *Handbook of Mathematical Functions* (Dover Publications Inc, New York, 1972)
11. A.H. Nuttall, IEEE Trans. Acoust. Speech Signal Process. **ASSP-29**, 84–91 (1981)
12. J.-P. Thiran, IEEE Trans. Circuit Theory **CT-18**, 659–664 (1971)
13. A. Fettweis, IEEE Trans. Audio Electroacoust. **AU-20**, 112–114 (1972)
14. V. Välimäki, Helsinki University of Technology, Faculty of Electrical Engineering, Laboratory of Acoustics and Audio Signal Processing, Report 37, ISBN 951-22-2880-7, ISSN 0356-083X (1995)
15. W. Gröbner, N. Hofreiter, *Integraltafel, Zweiter Teil: Bestimmte Integrale* (Springer, Wien, 1961)
16. W. Gröbner, N. Hofreiter, *Integraltafel, Erster Teil: Unbestimmte Integrale* (Springer, Wien, 1965)

© Springer International Publishing Switzerland 2015
T. Butz, *Fourier Transformation for Pedestrians*, Undergraduate Lecture
Notes in Physics, DOI 10.1007/978-3-319-16985-9

Index

© Springer International Publishing Switzerland 2015
T. Butz, *Fourier Transformation for Pedestrians*, Undergraduate Lecture
Notes in Physics, DOI 10.1007/978-3-319-16985-9

40869475R00146

Made in the USA
Middletown, DE
01 April 2019